URBAN LEGENDS

- FROM -

SPACE

The Biggest Myths About Space
✦ Demystified ✦

BOB KING

Author of *Wonders of the Night Sky* and *Night Sky with the Naked Eye*

PAGE STREET
PUBLISHING CO.

PAGE STREET
PUBLISHING CO.

First published in 2019 by
Page Street Publishing Co.
27 Congress Street, Suite 105
Salem, MA 01970
www.pagestreetpublishing.com

Distributed by Macmillan, sales in Canada by The Canadian Manda Group.

23   22   21   20   19   1   2   3   4   5

ISBN-13: 978-1-62414-896-5
ISBN-10: 1-62414-896-4

Library of Congress Control Number: 2019940333

Cover and book design by Laura Benton for Page Street Publishing Co.

Photography by Bob King, Richard Klawitter, William Wiethoff, Ondrej Králik, Jaye White, Todd R. Pederson, Svend Buhl, Frank Barrett, Vidur Parkash, Gianluca Masi, Alain Wong, Michael Kleiman, Aleksandr Ivanov, Jim Campbell, CPG 100, Peter J. Restivo, Malin Space Science Systems, Southwest Research Institute, Stellarium, Associated Press, Neal Herbert, ESO, Event Horizon Telescope Collaboration, ESA, NASA, JAXA and Hubble Heritage Team.

Graphics by Gary Meader, Bob King, Tom Ruen, Tim Pyle/Caltech-MIT-LIGO Laboratory, Xaonon, Orion8, Peris Lagios 1999, Kvr.lohith, Alan Chamberlin-JPL-Caltech, Lsmpascal, Amit 6, Flat Earth Society, Starry Night Education, NASA-CXC-M. Weiss, ESO-A. Roquette and ESO-M.Kornmess.

Printed and bound in the United States

DEDICATION

To journalists everywhere who seek the truth
and share it.

CONTENTS

PLANETS, COMETS AND ASTEROIDS 117

SUN, STARS AND SPACE 173

INTRODUCTION

+ ✦ +

Did we land on the moon? Is the Earth flat? Can you believe we're still asking ourselves these questions? If you're sometimes confused about what you read or see on the Internet, you're not alone. There's more information available at our fingertips than ever before, but telling fact from fiction isn't always easy. Misinformation about space topics seems to travel faster than light thanks to social media, making it difficult to know what to believe and what not to. Lots of things that *sound* true simply aren't upon closer inspection.

In this book, we'll take a look at common misconceptions, faulty science and wild claims. In the process, we'll explore lots of different aspects of my favorite hobby, astronomy. My aim is not so much to point fingers as to crack open a claim and let the facts speak for themselves. I want you to walk away armed with information and the necessary tools to distinguish fact from fancy. Future generations depend on us being informed citizens; my hope is that this book will throw some light on that path.

At the heart of seeking clarity in a blurry world of competing claims is the act of observation—becoming a keen observer of the world around you. All of us have this ability. By paying attention to nature and the voice of our inner skeptic, we can easily explain some of the manufactured mysteries like "second suns" and "chemtrails." You wouldn't buy a car, a home or even a mobile phone without doing the research, right?

The Scientific Method

How do you establish a fact? The scientific method has served us well for centuries when it comes to understanding the natural world. It's deceptively simple but has led to great discoveries that have benefited humanity in countless ways. Here's how it works in a nutshell:

1. Make an observation.

2. Ask a question.

3. Propose a hypothesis.

4. Make a prediction.

5. Test your prediction by experiment.

6. Check to see if the hypothesis was correct. If not, come up with a new one and devise another experiment.

If your results jive with your hypothesis, then the next step is to share the data with other scientists in your field so they can check your results with their own experiments. If they confirm your hypothesis, you may have made an important discovery. But if no one can replicate your results, then it's time to go back to the drawing board. *Repeatability* is one of the most important facets of the scientific method. Anyone can make a claim, but if no one else can reproduce it, your new idea is unlikely to be taken seriously.

Using the scientific method, scientists have vastly increased our knowledge of the world—from viruses to galaxies. In more practical terms, science has brought us lifesaving drugs, sanitation procedures that have greatly reduced disease and GPS satellites that help us find the nearest ice cream shop.

You tap into the scientific way of thinking when buying a car by studying the different cars available, comparing your favorites against similar models and taking one or more test drives. Many of us consult outside experts and online resources. Based on the data, you make an informed choice. This book will encourage you to apply the same kind of scrutiny to the increasing number of dubious and false claims often purveyed on the Internet.

The scientific method is one of most useful tools ever developed by humans. At its heart is simple curiosity, something we all possess from the moment we're born. Then comes the hard work to reach a conclusion based on evidence. That's why when I hear of a new "theory" about comets or the Earth being flat, I politely expect that the scientific method be used to verify and confirm the truth of the observations. And when it comes to radical proposals involving physics and astronomy, we should also expect mathematical underpinning to accompany the experiments and computer models that support the observations. Finally, the work should be published in a peer-reviewed journal so it can be shared and verified.

If someone skips any of those steps, I'm instantly skeptical. You should be too. I'm even more skeptical if the proposal runs counter to well-established scientific fact, like the Earth being flat or stars that run on electricity. Sometimes we're seduced by explanations that play into our prejudices. It may just *feel right*, appeal to our rebellious nature against "the man" or feed a worldview that sees the government as nothing but nefarious. As understandable as these viewpoints are, they don't justify replacing imperfect but *real* science with catchall *junk* science. New "theories" that run counter to established fact should be viewed as unconfirmed at best and false at worst. Experiment, hypothesize, analyze, verify, share and publish—then get back to us.

Science sets a high bar. This can frustrate both working and wannabe scientists alike. Scientists don't necessarily change their minds on the turn of a dime. Many are reticent even when the proof is crystal clear and will defend an idea to the death. This is understandable. Traditional science is skeptical of radical new views—as it should be. The burden of proof is on the experimenter and their new hypothesis. But if the data is good, makes accurate predictions and can be duplicated, eventually tradition acquiesces, and the new concept is welcomed into the fold. In this way, science evolves and changes.

If someone's new theory doesn't do those things, it may have its adherents but lack a factual backbone. That said, science has its limitations. For one thing, it has no *final* answers. All explanations are tentative. Understandably, that can be troubling. Most of us like a definitive black-and-white answer, but science is a subtle blade that forever finds yet another layer in the onion. Discovering something new is both a joy and a frustration because a fresh nugget of knowledge ruffles the current order of things. This ever-evolving aspect of science challenges how we see the world. I like to see it as a way to clear the cobwebs from our heads!

Some things that were once considered fact, like a medium in space called *ether* for light to travel by, was ultimately proved both nonexistent and unnecessary. I'm sure a few scientists fought like mad to keep the ether alive. It's gone now. In its place we have a better explanation for light's behavior.

Science builds upon a foundation of previous discovery. Einstein's theory of relativity superseded Newton's law of gravity because it provided a more comprehensive explanation of the concept of gravity. Not that Newton's law was incorrect. It just didn't work in certain circumstances like Einstein's does. To this day, scientists still use Newton's laws when talking about things traveling at well below the speed of light. Since most things do, Sir Isaac's laws will be around for a good long time.

Science has other limitations. It can't prove or disprove the existence of God or any other supernatural entity. It's *super*natural, outside of nature, and therefore beyond experiment. By the same token, intelligent design—the idea that a divine power must exist based on the incredible complexity of living organisms—is not science because it presumes an entity that's untestable.

Despite its flaws, the scientific method and viewpoint is the only one I know of that can get us out of almost any fix *and* offer plausible explanations for seemingly inexplicable natural phenomena. Science finds patterns that join things together in ways we could have never imagined. Did you know that trees in a forest "talk" to each other using root fungi as a biological version of the Internet?

Becoming a Good Observer

Some people are excellent observers of wildlife. Others notice the smallest details of a leaf or what someone is wearing. My specialty is watching the sky, but all nature fascinates. I've discovered that the closer attention I pay to some aspect of nature, the more nature reveals. As a boy, I loved watching clouds and still do to this day. Now I can distinguish between different cloud types, appreciate rare cloud formations and even make modest weather predictions based on cloud sequences. Familiarity with our immediate environment not only creates a closer bond with the natural world but also stokes curiosity. Curiosity leads to questions, which lead to further study and a deeper familiarity with the subject and occasionally a brand-new discovery.

I'm reminded of the Alberta (Canada) Aurora Chasers group. In 2016 its members helped identify a strange, aurora-like feature called STEVE (Strong Thermal Emission Velocity Enhancement) few had ever paid attention to. Members took photos of the snaking light, which caught the attention of a scientist who looked into the phenomenon and discovered that it was unique and unrelated to the northern lights. Cool!

I used to think that chickadees had one call, a perky *chick-a-dee*, but came to realize through more focused listening that I was way off. Scientists have discovered at least fifteen different vocalizations. Combine observation with knowledge, and even the most ordinary things become sources of wonder. Shadows. Beetles. Moonlight. Mud. Can you think of any other enterprise that's equally serious about discovering our cosmic origins as it is about devising a way to pop popcorn with microwaves?

I believe we're born with the innate ability to notice our environment. Whatever you find engrossing about the world, throw yourself into it. Besides the joy of connecting, the more you know, the less likely you'll be fooled or confused by those who innocently or deliberately misidentify natural events as UFOs, "chemtrails," double suns, fires in the sky or a planet on a collision course with Earth. Canadian astronaut Chris Hadfield put it best: "The more you know, the less you fear."

How to Verify Information Using the Web

To aid in your quest for straight answers, here are a few sites you can bookmark to verify claims or get great answers to further questions you might have. May your curiosity lead you onward to personal discovery!

To find or check facts you need reliable sources. Here are several you'll find helpful:

- Khan Academy, the scientific method (https://bit.ly/2Q8j9z2): The scientific method in action.

- Snopes.com: Excellent debunking site. Do a topic search, check the archive or submit your own topic.

- Ask an Astronomer (curious.astro.cornell.edu): Wonderful source of expert information from beginner to advanced levels on all things astronomical. Submit a question of your own.

- RationalWiki (rationalwiki.org/wiki/pseudoscience): Critiques and challenges pseudoscience and anti-science claims.

- Google Scholar (scholar.google.com): Search for original scientific papers, books and other sources across the Web, universities and other sites.

- MediaWise/Poynter Institute (poynter.org/mediawise): Program aimed at helping teens discern fact from fiction online.

Want to become a citizen scientist and contribute to our knowledge of the universe? Anyone can do it. Visit Zooniverse (zooniverse.org) and jump right in.

EARTH

THE EARTH IS FLAT

+ + +

Our ancestors proved the Earth wasn't flat several thousand years ago, but some would have you believe it's all a conspiracy. So let's take some time to show again why it must be so. The most satisfying and convincing way would be for each one of you to take a rocket into space and have a look for yourself. One day that may be possible, but we've got a budget to stick to, so let's examine a few less expensive alternatives.

The ancient Greeks knew that the Earth was a sphere and proved it with nothing more than their eyes. You can too. Here's Aristotle writing about 350 B.C.E. about observing lunar eclipses:

"The earth is spherical . . . in eclipses the outline is always curved: and, since it is the interposition of the earth that makes the eclipse, the form of this line will be caused by the form of the earth's surface, which is therefore spherical."

Even if you've never seen a lunar eclipse you can look at photos and videos of eclipses online and see the arc of Earth's shadow darken the moon. But wait. Wouldn't a hockey puck-shaped Earth cast a curved shadow? It would! So maybe those who believe in a flat Earth are right after all. Well, no. Their version of Earth is level, not tipped up on its side. Strictly speaking you can't prove Earth's a sphere by the shadow it casts during an eclipse from a single location, but informed by the hundreds of thousands of photos taken from orbit by astronauts and satellites since the dawn of the Space Age, we can see with our own eyes that we live on a globe without narrow, puck-like edges and certainly not the pizza-flat Earth envisioned by the Flat Earth Society. What's on the other side of that pizza anyway?

There are many more clues we live on a sphere. On a spherical planet, the force of gravity is the same everywhere because balls of matter "pull" their stuff towards their centers. On a flat Earth, the center of the "sheet" would have more pull, so you'd weigh more there than along its edge. That would mean a person traveling from the North Pole — the center of a flat Earth — would gradually lose weight as they traveled south to Antarctica, which defines its outer rim. No evidence for this exists.

More consequentially, if we lived on a flat Earth, the sun would always be present in the sky with nowhere to go at night. Flat Earthers get around this by claiming that the pancake-like Earth sits at the center of the solar system with the sun shining down upon it like a spotlight, illuminating only a portion of the landscape at a time (see diagram on page 16). There are at least two problems with this: the sun, not the Earth, sits at the center of the solar system, a fact decisively proven long ago, and it doesn't shine narrowly like a spotlight because it's a sphere that radiates from every single pore of its surface. No blinders or shutters aim and direct sunlight to one part of the Earth at a time.

Instead, the sun shines on one half of the spherical Earth, while the side opposite the sun experiences night. As the planet spins, what was once in darkness turns toward the sunlight, while the sunny side returns to night. It's Earth's rotation that makes the sun appear to move across the sky from sunrise till sunset. So much easier!

The sun is 93 million miles (150 million km) away—an enormous distance compared to Earth's tiny size. On a flat Earth it would appear in virtually the same spot of the sky no matter where you stood because of something called parallax.

Parallax is the apparent shift in the position of an object when seen from two different lines of sight. It's easy to understand if you hold an index finger up a few inches in front of your face as you alternately open and close your right and left eyes. Go ahead, give it a try. See how your finger jumps from side to side depending on which eye is open? That's a lot of parallax because your finger is so close compared to the distance between your eyes. But if you hold it at arm's length—about twelve times the distance between your eyeballs—it shifts much less. Next, imagine blinking at a distant mountaintop 10 miles (16 km) away. Blink as much as you want but the mountain will stay put because it's SO much farther than the distance between your eyes.

But if you could increase the space between your eyes to several miles, you'd see the mountaintop shift. What if you could expand the eye-to-eye distance to 7,926 miles (12,756 km), the diameter of the Earth whether round or flat? Wouldn't we be able to see a noticeable shift in the sun? No. The sun is so many millions of miles away that the amount of shift from one edge of a flat Earth to the other would be only a tiny fraction of a degree, too small to discern with human vision. And that means that no matter where you stood on a flat Earth at say, high noon, the sun would occupy virtually the same spot in the sky—that is, it would be high noon everywhere. For it to appear in different places in the sky, it would have to have a monster parallax and thereby so close that it would quickly vaporize the planet.

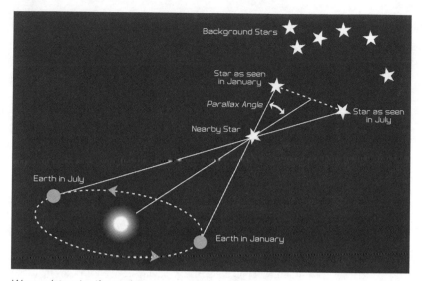

We can determine if something is near or far by measuring its apparent shift against the background stars from two widely separated viewpoints. If the shift or parallax is small, as it is for the sun, then the object is very far away. (Starry Night Education)

It would make me very sad if Earth were alone in being a flat world. Have you ever looked through a telescope at a planet? Some like Mercury are small spheres; others like Jupiter are giant spheres. But they are all massive enough that the material of which they're composed pulls itself into a ball through the attractive force of gravity. They can't help it—laws of physics make it happen. Small objects can be lots of shapes because they're not massive enough to scrunch themselves into balls, but massive objects have what it takes to crush in on themselves. So, if the little moon—only a quarter the size of Earth—can be a sphere, then doggone it, so can Earth.

Flat Earthers recognize spherical planets but set the Earth apart from them because of its special nature and position at the center of the solar system. It's understandable to call out our planet as unique. It is, after all, the only one we know of with life, but life or no, Earth doesn't get a pass from physics. We're not *that* different.

Here on the ground we find other ways to know we live on a sphere. If you visit a large lake or spend time by the ocean, you'll notice that you can only see the tops of distant ships. Their bottoms are temporarily invisible because they're hidden by the curvature of the Earth. As the ship heads to port, its bottom slowly rises into view. If the Earth were flat, a distant ship would appear very, very tiny to

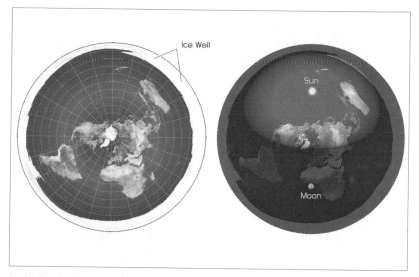

In the Flat Earth Model the Earth is pizza-pie flat and bordered by an ice wall. The sun (top) and moon are nearby spotlight-like light sources that orbit clockwise in a circle above the Earth, which is located at the center of the solar system. (Flat Earth Society)

be sure, but its entire shape would always be visible. Flat Earthers will often say that even from a high-flying airplane, the Earth appears flat. They are correct. While it might be possible to see a hint of the Earth's curvature at the altitude of a transcontinental jet under ideal circumstances, it's plainly visible from the space station.

A typical house dust mite is 0.012 inch (0.3 mm) across. From its perspective, the curve of a beach ball looks perfectly flat. If you tweeze the mite and hold it a fraction of an inch above the beach ball, similar to our perspective in an airplane above the Earth, the ball would still look flat. That's the situation we find ourselves in—mites aboard the mighty Earth.

Allow me just one more minute of your time for one of my favorite proofs. From my home in northeastern Minnesota, I will never see the second brightest nighttime star in the sky, Canopus in the constellation Carina the Keel. That's because the star never clears my southern horizon. The only way to see it is to travel south. Care to join me for a ride? About the time we'd hit Little Rock, Arkansas, Canopus would make its first appearance in the South. If we kept going, the star would rise higher and higher until at the southern tip of South America it would shine from directly overhead.

Let's keep going south to Antarctica and from there turn back north to Minnesota. As we do, Canopus sinks back into the southern sky and then drops below the horizon, disappearing from view once we're north of Little Rock again. This kind of behavior—the star, not the insane drive—is only possible on a sphere. If Earth were a flat surface as big as its diameter, then Canopus would *always* be visible no matter where you stood. It would also be at exactly the same altitude wherever you were on that sheet. Why? Because it's so far away—1.8 quintillion miles (310 light-years)—compared to the Earth's size that no amount of travel would cause it to budge from its position in the sky. For the star to appear to move up or down in the sky as you travel by car or plane on a flat surface, it would have to be extremely close to the Earth, just like the sun in our earlier example. So close it would—dare I say it again?—vaporize the planet.

Lots of us find the beliefs of Flat Earthers curious because they run counter to long-accepted fact. It's so easy to show that we live on a sphere. Why go through contortions to prove otherwise? Humans love to be contrary. When I was younger, my parents tried to guide me toward making good choices. Did I follow their wisdom? Sometimes, but I often rejected those values and did it my own way. Maybe you were like me.

Rebellion is part of our nature. It starts when we're babies and continues in one form or another throughout our lives. I believe this rebellious spirit permeates fringe groups like the Flat Earth Society. I understand the appeal of being an iconoclast and railing against "the man" of big science. Combined with the mind-set that "my opinion counts too," people find a common bond through shared beliefs, whether those beliefs have a basis in fact or not.

When it comes to the flat Earth concept, you might ask what harm there is in believing the idea. Beware the slippery slope! If you start walling off real science for the tribal version, you're liable to believe in anything as long as it "feels" right. Ignorance—sometimes willful—and confusion fuel the anti-vaccination movement, where a rejection of scientific studies can have lethal results.

We're all entitled to our opinion, but I'm more likely to believe you or be swayed to change my opinion if yours has a basis in fact. While we might prefer to believe in an alternative reality, facts are stubborn things. You can toss them aside, but beware—they will come back to bite you in the end. As Katharine Hayhoe, a climate scientist at Texas Technical University, noted: "Facts aren't something we need to believe in to make them true."

CONTRAILS ARE REALLY CHEMTRAILS

If you've ever looked up and spied a long, chalk-like streak across the blue sky, you've seen a condensation trail or *contrail*. If it's still sharp and narrow, you can follow it back to its source at the tail of an airplane. Contrails are often seen trailing a short distance behind high-flying aircraft, typically commercial jetliners making cross-country flights.

Contrails are narrow streams of clouds that form behind each of a plane's jet engines when water condenses on tiny, soot-like exhaust particles produced by the burning of jet fuel. Most of the water comes from the surrounding air, with a smaller share from the plane's engines. They frequently form at high altitudes from about 28,000 to 40,000 feet (8.5 to 12 km), where the air is extremely cold, often below -40°F (-40°C). That's why the planes that create them often look so tiny—they're far away. Depending on humidity and wind direction at airplane altitudes, a contrail can disappear quickly or linger and spread into a band of thin clouds.

Closer-to-home examples of contrail-like phenomena include seeing your breath on a cold day and the snaking plumes of tailpipe exhaust from our cars in arctic weather. In both cases, water-vapor-laden air exits a warm place and enters a cold place, where it quickly condenses into fog.

People have been watching airplane contrails since the early days of aviation. A fellow named Ettenreich reported seeing the first contrail in 1915 over the Italian Alps, describing it as "the condensation of a cumulus stripe from the exhaust gases of an aircraft." During World War II bombing runs, contrails could hurt or hinder a mission. When hundreds of planes were involved, it was difficult to stay in formation or sight a target. On the plus side, the spreading stripes of multiple trails provided cover from the enemy.

With 87,000 flights crisscrossing the United States alone every day, silvery jets and their white contrails have become a common sight in our time. Because

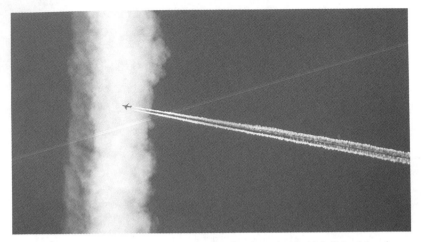

An older contrail (left) and a fresh contrail created in the wake of a high-flying jet make a striking pattern in the sky. Contrails form when water vapor in engine exhaust condenses into clouds in cold air at high altitudes. (Bob King)

aircrafts fly a variety of paths to destinations, their contrails often cross and intersect to make interesting patterns. I always keep my mobile phone at the ready because you never know when you'll see something striking. Contrail watching is an enjoyable activity that helps us become better observers. Next time you're planning an outdoor activity, I encourage you to keep an eye out for them. While man-made, they're nature enhanced and reveal much about the state of the atmosphere overhead.

Binoculars will reveal fascinating loops and whorls within the finger-wide trails caused by turbulence as the exhaust leaves the plane's engines. Also look for the gap between the engine and the start of the contrail. Typically about 100 feet (30 m) long, it represents the amount of time it takes for the water in the exhaust to freeze into cloud droplets in the cold air. At 36,000 feet (12 km), the typical air temperature hovers around -70°F (-57°C).

There are three basic kinds of contrails: short-lived, persistent (non-spreading) and persistent spreading. The short-lived ones look like short white lines closely following the planes. They form only when the amount of water vapor present at the plane's altitude is low. The trail disappears quickly because the ice that condenses onto the exhaust particles quickly turns back to vapor in the dry atmosphere. If you've ever seen a contrail that looks like a series of dashes with blank spaces in between, the plane is encountering "dry patches" of air as it zooms along.

People make contrails, too! On cold days, the water in our warm breath condenses to form clouds of vapor similar to how contrails form. (Alain Wong / CC 0 1.0 Wikimedia)

Persistent, non-spreading contrails also look like narrow streaks, but they remain in the sky after the plane has passed. They indicate more humid air with lots of water vapor available to form a long-lasting contrail. Persistent spreading contrails likewise form in a humid atmosphere but can spread outward to form a layer of cirrus clouds depending upon temperature and local winds at the plane's altitude.

When multiple planes pass over and produce spreading contrails, the trails can sometimes expand and merge into a hazy, nearly overcast sky. If you live under a busy flyway, you may have seen this yourself. There's nothing sinister about contrails, despite what you might hear online about planes spraying chemicals on the population as part of a secret government experiment. These so-called "chemtrails" are said to be toxic, but if that's true, then the indiscriminate spraying of them would have no target. They'd drift down over rich and poor alike. And to what purpose? No scientific study would be done this way because there's no way to control it. Not to mention that thousands of airline pilots and support staff would have to be secretly in on the gig. We've long understood contrails to be the result of natural processes at work on human-made pollutants.

That's not to say that chemicals haven't ever been released from high-flying planes. Cloud seeding using materials such as silver iodide, salt and dry ice has been underway since the late 1940s with the goal of increasing precipitation in dry regions. The materials act like seeds to assist the natural condensation of water vapor, encouraging the formation of clouds and (hopefully) rain. Results

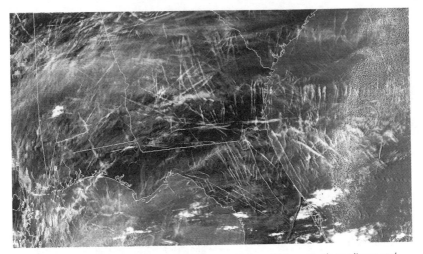

In busy flyways when atmospheric conditions are right, multiple contrails can linger and expand, clouding the sky as shown in this satellite image. (NASA Langley Research Center)

have been mixed—some experiments appear to have increased rain and snowfall, others have not. Weather modification through seeding remains hit and miss.

Some people have rightly expressed concern about the effect of silver iodide on people and the environment. Exposure to excess silver can result in a form of skin discoloration called *argyria*. But studies have shown that while precipitation connected to cloud seeding does increase the amount of silver in drinking water, it's far below the 100 parts per billion clean-drinking water standard for silver. Whether you consider cloud seeding a potential problem or not, commercial aircraft, the ones we see crisscross the sky every day, aren't the ones doing the seeding. Seeding is handled primarily by private enterprise using smaller planes and specialized equipment over a selected area at a particular time.

Do contrails have other affects? Meteorologists will tell you they have concerns about spreading contrails—the ones that form clouds—because they cause an increase in cloudiness, which could have an effect on climate. Contrails reflect sunlight and can lead to a slight cooling effect, but they also, like any cloud, trap heat rising from the ground below, so it's hard to say which effect dominates. Right now, the best hypothesis suggests that contrails contribute *slightly* to global warming.

Contrails are sometimes confused for other phenomena. I've seen photos claiming to be fireballs (a brilliant type of meteor) that are actually distant contrails lit

Seen from a distance especially around sunrise and sunset, short-lived contrails mimic comets and bright meteors. But their rapidly changing form soon reveals their true nature. (Bob King)

up red by the rising or setting sun. You can tell the difference between the two because nearly all meteors last only a few seconds, while contrails remain in sight for minutes. A meteor makes a single streak; contrails often show multiple, parallel streams. Binoculars are a great help in telling the two apart.

Contrails *can* also be confused with comets because of their similar appearance, especially when the plane is far away and the trail is compressed by distance. Once while driving, I spotted a brilliant white, comet-like object low in the western sky. Its appearance was so striking that I pulled the car over for a better look. It only took a minute to realize my comet was an airplane, its contrail compressed by distance and lit spectacularly by the sun.

Comets, even bright ones, are so far away that they appear to move much too slowly to see movement with the unaided eye, even after many minutes.

The closer we observe the world, the more familiar we become with what comes and goes and the more discriminating our powers of observation become.

Resources
- Silver iodide cloud seeding: http://health.utah.gov/enviroepi/appletree/technicalassists/Silver_Iodide_Cloud_Seeding.pdf

UFOS ARE REAL

+ ✦ +

When I teach community astronomy classes, I'll sometimes ask if anyone has seen a UFO. Hands always go up. When the person describes their experience, I can sometimes tell what it might have been and sometimes not. I need more information. But I always wish I could have been standing next to them at the time and have seen it for myself, for two reasons. First, maybe I could have explained the phenomenon based on long experience with the night and its crazy lights. Second, despite having surveyed the sky for more than 4,000 nights over the past 50-plus years with eye, binoculars and telescope, I've yet to see a UFO. I'm starting to think aliens don't like me.

OK, when I was about thirteen, I saw a V-shaped formation of moving lights through my telescope that had me panting for a moment until I realized it belonged to earthly geese illuminated by Chicago light pollution. Back then I was primed and ready for UFOs. I used to hang out in the magazine section at the local Walgreens, paging through the monthly UFO magazines popular in the mid- to late 1960s. After so many years of skywatching, I can't say I've seen anything that didn't have a natural explanation.

But how could 104,797 people be wrong? That's the number of UFO sightings reported across the United States since about the mid-1930s through early February 2019. All the observations are kept on file at the National UFO Reporting Center (NUFORC). Of that tally there are undoubtedly multiple reports from the same people, but the number still astounds. And that's just the United States. Worldwide, thousands of additional sightings are described.

The reports remained a trickle from the 1930s through the 1980s but grew to a torrent by the mid-1990s. Numbers per state are related to population and possibly the presence of rocket launch sites. California, the most populous state and home to two rocket bases, came out far ahead of all the others with 13,480 sightings. Florida took second place with 6,459 and Washington State third with 5,870. Florida has Cape Canaveral, where it might be easy to confuse rocket launch activity with UFOs, but Washington?

This object, submitted as a photograph of a UFO, is actually a reflection of the sun inside the mobile phone's camera lenses. Such "internal reflections" occur when the camera is pointed toward a brilliant light source, in this case, the sun. It's easy to recreate — give it a try with your own phone. (CC BY SA 4.0 Wikimedia)

If you read the reports, you'll discover that nearly all observers describe some sort of light phenomenon in the sky: red orbs, blue orbs, green ovals, blinking lights, rotating "washers," metallic spheres and lights flying in formation.

The upside of all these reports is that people are looking up and noticing things in the sky. But one has to ask whether any of these sightings represent an actual spacecraft piloted by extraterrestrials. That's a high bar requiring A LOT of hard evidence. What looks like a strange oval "glowing from within" to you is an isolated patch of the aurora to me. A bright, sausage-shaped light may be the next person's meteor. That glinting, metallic ball with a vapor trail is really a silver Mylar balloon (with string attached) that a child accidentally let go of. That last one appears in a documented report. After I saw the blurry video, there was no question of what it was.

Things look like other things, and unless we can prove otherwise, a light in the sky—especially the thousands of varied lights that are reported each year as UFOs—can be many things, the least likely of which are aliens. Most people don't pay a lot of attention to the sky. Who can blame them? Their lives are busy enough. The occasional moments we do look up and see something we can't explain, it's natural to consider an extraterrestrial origin. You're even more likely to ascribe a strange phenomenon to an ET if you're predisposed to think UFOs are

Red sprites frolic above a thunderstorm over southeastern Germany just before midnight on August 18, 2017. At the time, the storm was about 315 miles (507 km) away. The display lasted only a fraction of a second. (Ondrej Králik)

real and alien visits a fact. What we bring to our observation of the natural world is colored by what we believe and reinforced if we share those beliefs with a group of like-minded friends and acquaintances.

While it may lack the storytelling power of an alien ship swooshing by, those unusual lights in the sky often have natural causes, many of them fascinating in their own right. That doesn't mean there aren't unknown natural phenomena that require persistent observation to confirm and verify. A good example are sprites, the red, jellyfish-shaped electrical discharges that appear high above powerful thunderstorms. Scientists were initially skeptical of airplane pilots' reports of seeing them, but now we know they're real. Equipped with off-the-shelf digital cameras, people are routinely photographing these gangly, evanescent flashes from the ground.

The argument that UFOs are extraterrestrial spacecraft and that humans have routinely encountered aliens—mostly while being experimented on—fails in several areas. Despite hundreds of thousands of sightings, there has been no scientifically verified contact or communication with aliens. No sharing of technology. No communication channels opened to stay in touch and share ideas. No alien artifacts. You'll read about smells, burn marks, flattened fields, but not a single verified piece of alien technology has been produced, studied and written up in a peer-reviewed scientific publication.

Moreover, the Pentagon has undertaken two massive studies of UFOs and UAPs (unidentified aerial phenomena) to address sightings by citizens and the military, particularly pilots. Neither Project Bluebook (1947 to 1969) nor the more recent Advanced Aviation Threat Identification Program, or AATIP (2007 to 2012), produced any definitive results. No smoking guns. No UFO parts. AATIP received $22 million to investigate UFO reports and also fund research looking into speculative science like warp drives and invisibility cloaking.

The biggest reveal for the money spent? Several videos taken by U.S. Navy pilots in the early 2000s showing fuzzy specks flying at apparently impossible speeds. Should we count that as positive proof that alien beings are yanking our chain? While interesting and worth investigating further, they hardly rank as definitive evidence. In hopes of getting to the bottom of the unidentified aerial phenomena, the Navy drafted new guidelines in 2019 that encourage pilots to report their sightings without the fear of being stigmatized. Kudos to the Navy. I'm just as curious as you are to their origin. Let some air in, gather unbiased evidence and maybe we'll get an answer.

Can you imagine how the first definitive proof of extraterrestrial life would rock our beliefs to the core? To know we're not alone in this great big cosmos? Surely worth a Nobel Prize for the discoverer. Yet we continue to wait because to date no one has shown us the money.

Given the incredible diversity of supposed alien ships sighted in recent years, the number of alien species visiting from diverse planets would seem to be huge. If true, then why does nearly every sketch or purported photograph of a supposed ET bear striking similarities? They're all surprisingly humanoid, though often smaller and more childlike. Many have large, almond-shaped eyes, slits for mouths and no ears. If they really do come from other planets around faraway stars, they must have also evolved from simpler forms just like we did. Vision, moving about on limbs, fins and legs and hearing (among other senses) may be universal no matter evolution's diverse paths on different planets, but the likelihood that spacefaring, intelligent life would resemble smaller versions of us must be astronomically small.

Some people will say the government is covering up the truth about aliens, but I doubt it. I'm skeptical of the government too. But the world has lots of smart people with access to big audiences through the Internet. If I had incontrovertible evidence of aliens, such as a bona fide piece of their technology, richly detailed images and other recordings that fully documented my encounter, I would immediately let the cat out of the bag and arrange a meeting with a physics professor at the local university. Then we'd go from there. The last people to hear from me would be the "government."

Naturally, no one would believe my story at first, but after a thorough analysis of the alien artifact and the documentation accompanying my encounter, a few scientists across varied disciplines might conclude that my claim had some validity. Again, the technology would be the clincher. The recordings, photos, etc. would be treated as corroborating evidence. Following peer-reviewed published studies, my encounter might be considered *possible* but subject to further verification from repeat ET visits.

Scientific study is the best way to prove aliens exist because scientists scrutinize the evidence and test stuff over and over until they're satisfied. They have the knowledge and equipment to make definitive analyses. Scientists can be pretty hard-nosed, too, exactly what you'd want when it comes to proving something as incredible as ET visiting Earth in spaceships.

My scenario is only a mental exercise and will almost certainly not happen in my lifetime. But I hold out hope for at least some kind of contact through SETI (Search for Extraterrestrial Intelligence) programs like SETI@Home, Breakthrough Listen and Project Argus that listen and watch for artificial signals from outer space that could indicate an advanced intelligence.

I'd love to see a UFO, and I passionately believe there are other forms of life in the universe, everything from bacteria-like organisms to highly intelligent beings, but so many sightings and abductions strain credulity. Think how much time had to pass before Earth produced a species capable of conceiving and building the first crewed spaceship? Try 4.5 billion years. For 3 billion of those years, life was microscopic!

Many UFO phenomena can be explained as the observer's unfamiliarity with night sky phenomena like star twinkling, satellites or even the planet Venus. Some may have to do with dreams that seem as real as waking life, while others are wishful thinking. The vast spaces that separate us from even the nearest stars preclude easy coming and going by spaceships. Even our fastest spacecraft would require 50,000 years to reach the nearest star system, Alpha Centauri. You can argue that aliens have invented technology to travel huge distances in mere seconds.

Sometimes the aurora appears as isolated saucer-shaped patches of pulsating light that some might mistake for a UFO. (Bob King)

Maybe. Why not arrange a meeting, talk and share ideas and technology after such an incredible journey? Maybe we're too violent. Maybe they're afraid of being wiped out by Earth germs. Maybe interstellar law declares observation only, no contact. The fact that we can make up all kinds of things about what aliens do and don't do proves we know nothing about them. And we know nothing because, chances are, none have yet come to visit.

So what are all these things people are seeing and claiming as UFOs? Here are some of the most commonly mistaken natural objects thought to be extraterrestrials on the prowl:

- Venus or another bright planet like Jupiter or Mars. You might look up and notice Venus for the first time and think: "I never saw that before. Could it be a UFO?"

- Satellites. There are lots, and they travel in a variety of directions as well as flare and flash. Again, if you're not familiar with their appearance, you might mistake them for UFOs. You can identify which satellites are passing over your area at Heavens-above.com. Select your city and then click the "Daily Predictions for Brighter Satellites" link.

An inferior mirage over Lake Superior in Minnesota transforms an offshore island into a hovering "UFO." (Jaye White / Cascade Vacation Rentals)

- Weather balloons. They look silvery in sunlight and dark against a cloudy sky. Most drift steadily in one direction, but high-altitude balloons can linger over an area for hours. Occasionally, they get caught in updrafts and move quickly or erratically.

- Chinese candle lanterns set aloft. A single lantern looks like a "red star" slowly moving in one direction.

- Aurora. Most of us can distinguish the arcs and rays of a classical aurora, but sometimes an isolated patch can appear totally on its own and slowly pulsate for many minutes. Eerie, but definitely not caused by aliens.

- Rocket launches, especially the boosters that can leave colorful trails of condensation. Sometimes boosters vent fuel, and if they're spinning, it can create a spiral glow in the sky.

- Sirius, the brightest star and the one that twinkles strongest. Twinkling is often mistaken for an actual moving object.

- Bright meteors and comets. Comets don't usually change radically in a short period of time, but meteors can break up, explode or have twisted "tails."

- Lenticular clouds. They're often shaped like "flying saucers."

- Reflections of lighting fixtures off windows or glass walls mistakenly assumed to be outside and in the sky.

- Mirages caused by light traveling through air layers of different temperatures can create bizarre, hovering forms over roads and lakes that seem to defy gravity.

- Flares dropped by military aircraft during training missions.

- Flashing lights on airplanes. You're probably already familiar with the red and green flashing navigation lights on aircraft, but non-civilian planes often include different lighting configurations. During the writing of this book, I saw just such a plane that at first glance looked like a small fleet of UFOs—four small ones and two bigger. For a moment I was confused, until it banked and I could clearly see it was a plane.

- Lens flares and internal reflections in cameras and camera phones. If you didn't see it with your eyes, but it's on your mobile phone, that's lens flare. If you point your phone to another part of the sky, the reflection will move with it.

- Experimental military aircraft.

- The International Space Station (it's bright!) and flares from sunlight randomly reflected off satellites.

- Mylar party balloons (the shiny, silver kind). They can tip, twist and shimmer with multiple reflections when the sun is out.

- Moving stars. If you're not familiar with the bright stars and stare at one long enough—several minutes will do—it will appear to move. This is caused by Earth's rotation.

Consider one of these possibilities the next time you encounter a UFO. I'll make you a promise: the better you know the sky, the fewer UFOs you'll see. Sorry for the disappointing news. Knowledge is sometimes criticized for lessening the wonder of the world. I like to think that it instead deepens our understanding and increases our capacity for wonder. That's how it feels to me anyway. I love the "I never knew that" experience and suspect you do too. Knowledge also empowers us. With it, we're not helpless.

Maybe there are other reasons why believing in UFOs has become so pervasive in modern culture. At least a third of all Americans believe UFOs are extraterrestrial vehicles. Mystery, intrigue and "what ifs" play into our fascination. Aliens are often evasive. You'll never find one ordering soup from a deli. Their technology is light-years beyond our own, and they announce their presence with bizarre lights and other effects. We can't help but want to know more. You listen to your friend's story about seeing an ET or tune in to a news report and before long, you're a believer.

People also like a connection to the cosmos, whether that's through religious beliefs, science or even believing in UFOs. The universe is less lonely with aliens, and their superior technology gives them incredible powers far beyond the human ken. For some, they're akin to gods who care enough about us to check in and see how we're coming along.

What we believe becomes our truth. My fear is that when we believe in things without certain evidence, we lose our ability to distinguish fact from fancy and go astray. The world is not entirely a made-up thing. We may wish for events and expectations to turn out a certain way, but they don't always adhere to our desires.

Resources
- National UFO Reporting Center (NUFORC): nuforc.org

EARTH'S MAGNETIC POLES WILL FLIP AND ALL LIFE WILL PERISH

+ ✦ +

Ever used a compass? Its seemingly magical power to point north is sourced deep within the Earth 1,800 miles (2,890 km) beneath your feet. That's where the outer, molten-iron core of the planet begins. Churning like water boiling in a pot, and enhanced by our planet's rotation, electric currents are generated within the liquid iron, which creates a planet-wide magnetic field that aligns closely with the Earth's rotation axis. Your compass responds to the field—which has a north pole and a south pole, just like a classic bar magnet—by turning its needle north.

We rarely think about the connection between electricity and magnetism, but one begets the other. Try this experiment to see it with your own eyes. Get a D battery and a short length of wire. Place the compass under the wire and touch one end of the wire to one pole of the battery and the other end to the opposite pole. The electric current flowing along the wire generates a magnetic field that deflects the compass needle in a big way.

Earth's magnetic field reaches far into space and serves as a shield that helps lessen the severity of solar storms brought on by explosions called *flares* and *coronal mass ejections*, giant clouds of high-speed particles blasted into space by our fitful star. Like water off a duck's back, the field deflects much of what the sun flings our way, and we're no worse for the wear. Most of the time, anyway. When it doesn't, we experience a geomagnetic storm as particles hook into and follow the invisible lines of Earth's magnetic field down into the upper atmosphere to spark auroras and sometimes wreak havoc with satellite electronics. More on that in a minute.

So we're clear, a compass rarely points to the North Pole or true north. Instead, the needle aligns itself with Earth's magnetic field and indicates the direction of *magnetic* north. Due to irregularities in the flow of molten iron in the planet's outer core, magnetic north can differ from true north by as much as 30° west of north to 26° east of north. The difference is called *magnetic declination*. To determine true north, you must first look up the magnetic declination of the city or region before you start your journey and take the difference into account or you'll stray from your path.

Our magnetic defenses aren't static. We're used to compass needles pointing north, but 800,000 years ago north and south were exactly the opposite of where they are today. Every so often and completely unpredictably, Earth's magnetic field flips or reverses. South becomes north and north becomes south. There have been 183 reversals over the past 83 million years separated by intervals as short as thousands of years to 50 million years. For the past 20 million years, the average time between a pole reversal has been 200,000 to 300,000 years. Scientists are still trying to figure out what causes the flips, but it's thought that changes in the current flow of the ocean of iron under our feet is responsible.

Reversals are completely random and transition from one polarity to the next over a period of 1,000 to 10,000 years . . . usually. The last, called the Brunhes-Matuyama reversal, occurred 780,000 years ago and flipped in the space of a human lifetime. Flips aren't necessarily tidy affairs; sometimes multiple north and south poles form over different regions of the planet before settling into a single field.

Earth appears to be on the verge of the next reversal. The magnetic north pole, which is offset from the geographic pole, has been moving northward across Canada toward Siberia since at least the early nineteenth century and in recent years picked up speed, accelerating from 10 miles (16 km) per year to 40 miles (65 km) per year compared to a hundred years ago. But we must be careful to assume the hurry precludes a flip. What looks like an imminent reversal can suddenly peter out and quit, only to resume again later. While the current movement of the pole is telling, there's no guarantee the field will reverse anytime soon. Then again, it just might.

Should we worry? Because the magnetic field protects us from potentially dangerous solar storms and acts as a shield for cosmic rays (high-speed particles whirling around the galaxy), are we defenseless without it? During a reversal, Earth's magnetic defenses grow very weak but don't disappear entirely. Multiple

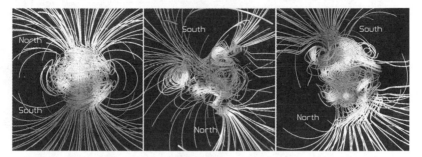

NASA computer simulation of a magnetic field reversal in progress. The tubes represent magnetic field lines. A field reversal can be a messy business with multiple magnetic poles before sorting itself out. (NASA)

magnetic poles may even form. But no matter what shape the magnetic field may assume, the atmosphere will help keep us safe from these microscopic missiles, absorbing and otherwise limiting the number of subatomic particles from the sun from reaching lower altitudes and the ground.

While not without defenses, we won't get off scot-free. Scientists predict the solar wind could enlarge current holes in the ozone layer and create new ones. Power grids could be disrupted unless properly hardened against solar storms. Birds, bacteria and other creatures that use the magnetic field to orient themselves will likely be confused, which could affect their ability to eat and migrate. Satellite electronics will also take a hit if not fully hardened against the deleterious effects of high-speed particles.

That said, we needn't be worried about the continents falling asunder and other such end-of-world scenarios. Beware the hype and remember that Earth has been through countless reversals. Studies of the fossil record during those uncertain times reveal no extinctions or anything else out of the ordinary. Some species undoubtedly suffered, but nothing like the great extinctions brought on by asteroid impacts and massive volcanic activity. One effect of a weakening magnetic field will find favor with future flippees—the possibility of seeing auroras as far south as the equator!

If you're wondering how we found out about magnetic reversals without having experienced one, let's take a trip to the bottom of the ocean. Magma oozes from great seams in Earth's crust called *mid-ocean ridges*, pushed up and over either side of the ridge, where it spreads, cools and solidifies into new crust. While still molten, iron grains within the lava orient themselves to the current magnetic field.

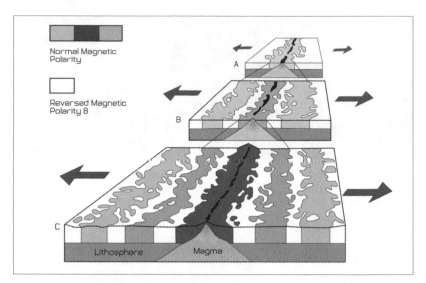

When new crust bubbles up as magma at mid-ocean ridges, it's stamped with the signature of the current magnetic field, creating a record of Earth's ever-changing magnetic field. (Public domain / Wikimedia)

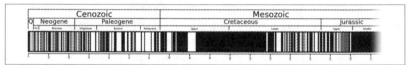

Dark areas in this time scale indicate periods where the polarity of Earth's magnetic field matched today's north and south magnetic poles, while light areas denote times of reversed polarity. The strip spans from the present time (left) to 170 million years ago. (Public domain / Wikimedia)

Once they cool, the field is essentially "stamped" into the rock. As long as the seam keeps producing lava, each magnetic reversal gets stamped into the fresh section of crust. The oldest crust—farthest from the spreading center—records the most ancient reversals. The closer to the ridge, the younger the rock, until at the ridge itself we arrive at the present day.

Scientists tow magnetometers across the ocean that can detect the strength and direction of the magnetic fields embedded in the rock, mapping our planet's past magnetic antics. Since material spreads on either side of the ridge, once the ship passes over the seam, we detect the same pattern in reverse on the other side—from young to old—as a series of stripes that looks a lot like a grocery store bar code.

I have to share a little secret about your compass. Because Earth's magnetic north and south poles are fairly closely aligned with its rotation axis, one end of a magnetized compass needle points north toward the North Pole and the other to the South. This is because opposite poles of two magnets attract—the north pole of one magnet snaps to the south pole of another. Likewise, similar poles repel: north pushes against north, south against south. Because Earth's *north* magnetic pole attracts the *north* end of the compass needle, it's *technically* the South Pole of our planet's magnetic field even though it's located in Canada! By convention, we call it north. Crazy, right?

Resources

- Compass deflection experiment instructions and diagram: https://how-things-work-science-projects.com/compass-deflection

- National Oceanic and Atmospheric Administration (NOAA) geomagnetism FAQ: https://www.timeanddate.com/geography/magnetic-declination.html

U.S. AURORAL RESEARCH CENTER "WEAPONIZES" THE WEATHER

+ + +

When was the last time you thought about the ionosphere? I bet it's been a while. But when it comes to solar storms and the aurora borealis, this upper layer of Earth's atmosphere is where the action is.

To study what's going on up there, the University of Alaska Fairbanks operates a research station known by the acronym HAARP (High Frequency Active Auroral Research Program) about 200 miles (322 km) northeast of Anchorage, outside the tiny town of Gakona, Alaska. Among other research carried on at the facility, HAARP uses a high-power transmitter focused through an array of 180 antennas to direct pulses of radio energy into the ionosphere to study its properties.

Before we plunge into what some consider a controversial project, let's first learn a little more about Earth's atmosphere.

All the weather and the air we breathe resides in the lowest layer, called the *troposphere*, which extends 11 miles (18 km) high over the equator and mid-latitudes but only 3.7 miles (6 km) over the polar regions in winter. Air is thickest here but thins and chills with altitude. Above it lies the stratosphere, a layer of stratified air that's colder at the bottom and warms with altitude. Warming occurs because ultraviolet light from the sun heats the ozone layer, which extends from the middle to the top of the stratosphere.

HAARP uses an array of 180 antennas to direct pulses of radio energy into the ionosphere to study its properties. (Michael Kleiman)

At about 31 miles (50 km) altitude, that layer gives way to the ionosphere, a deep but extremely rarified layer that reaches 620 miles (1,000 km) straight up into outer space. The ionosphere is made of ions, which are atoms and molecules of familiar gases like oxygen and nitrogen that have lost an electron. Ultraviolet and X-ray light from the sun are energetic enough to pry away electrons from these neutral gases, leaving them with a positive charge. In essence, they're electrified . . . but they don't stay that way for long. The atoms are all too eager to get back to normal, so they latch onto stray electrons in a process called *recombination*. Moments later, they're zapped again and the cycle begins anew.

There are several layers to the ionosphere, but the basic ones are labeled D (lowest), E (middle) and F (highest). Because of their electrical properties, some of these layers make excellent reflectors of radio waves during different times of the day and night as well as seasonally. Ham radio operators routinely bounce signals off the ionosphere to talk with other hams halfway around the world. Thanks to its reflective properties, shortwave radio enthusiasts in Peoria, Illinois, can listen to live newscasts from Papua New Guinea.

The ionosphere is also home to the aurora borealis. While the northern and southern lights can materialize in any of the three layers, the aurora does most of its dancing in the E layer. Protons and electrons shot out from the sun are guided by Earth's magnetic field into the ionosphere at extreme speeds. When they crash into oxygen and nitrogen atoms, they kick out electrons, temporarily converting them into ions. Moments later, the atoms recombine with electrons and release energy as shimmers and rays of green and red light.

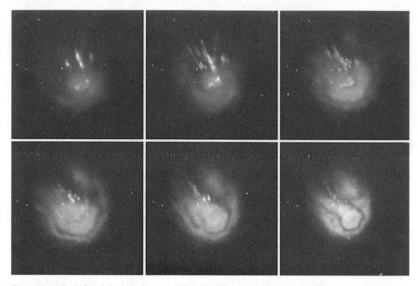

Examples of artificial aurora created with HAARP and visible with the naked eye. In color, these images show the classic pink and green color of the aurora. (Courtesy of Dr. Todd R. Pedersen)

Researchers at HAARP pay the University of Alaska to use the research station to work on projects ranging from bouncing radio signals off bits of ionosphere to communicating with submarines to creating artificial auroras to better understand the nature of the real ones. On March 10, 2004, scientists beamed timed pulses of high-frequency radio energy (5.95 MHz [megahertz], a frequency you can pick up on a shortwave radio) straight into an active E-layer aurora with the HAARP transmitter for 10 minutes with the hope of stimulating an artificial aurora.

It worked! Cameras recorded speckles of green light bright enough to see with the naked eye that were either produced by the transmitter beam directly or indirectly by modifying the rate aurora particles bombarded the atmosphere.

Any time you go shooting beams of energy into the atmosphere, people naturally wonder whether there may be possible side effects—especially if it's a government-run program, in this case the U.S. Navy and Air Force, which operated HAARP from its start in 1993 until the University of Alaska took over the facility in 2015. Warranted or not, people have a distrust of government that on the one hand can lead to greater transparency and on the other to clouding the truth through conspiracy making. With HAARP, the military's purpose was to learn more about the ionosphere and how to use it to affect radio communications over long distances.

Despite the fact that the station was then and continues to be open about its research, rumors began to circulate that scientists were using the facility to control the weather, shoot down satellites and even reach into our minds to control our thoughts. Experiments at HAARP were blamed for the 2011 earthquake and tsunami in Japan and a variety of other catastrophic events with well-understood *natural* causes.

Let's look at the facts. First, HAARP is not a classified or secret project. Scientists who have used the facility routinely publish their research as scientific papers that can be downloaded for a small fee (some are free) online. Second, the ionosphere is a dynamic medium, subject to the solar wind, flares and even lightning from thunderstorms. Heating a small bit of it at a specific altitude with radio waves is equivalent to dropping a stone into a pond. In a matter of seconds, the ripples created on the pond's surface dissipate with no lasting effects. Similarly, the energized bit of ionosphere reverts back to normal within seconds and at most several minutes. A single stroke of lightning possesses about 10 gigawatts of power on average, far more than what the radio transmitter can muster. The best HAARP can do is 3.6 megawatts of radio frequency power—absolutely pathetic next to the "1.21 *gigawatts*" needed to fire up Doc Brown's DeLorean time machine in the *Back to the Future* movie series. Just trying to keep the mood light here.

Nor can HAARP alter the weather. U.S. lightning detection systems monitor an average of 25 million cloud-to-ground lightning strikes a year from approximately 100,000 thunderstorms. If that doesn't change the weather, HAARP's measly megawatts can't either.

We live in a sea of radio waves. AM and FM radio stations transmit radio energy through air, buildings and automobiles so we can listen to news and music. Higher-frequency TV signals likewise saturate our environment, and yet you and I stroll through this invisible energy every day and never give it a moment's thought. Knowing a few basic scientific facts can keep fear and confusion at bay.

Resources

- HAARP main site and FAQ: https://www.gi.alaska.edu/facilities/haarp

- HAARP artificial aurora experiment: https://apps.dtic.mil/dtic/tr/fulltext/u2/a435730.pdf

THE GREAT WALL OF CHINA IS VISIBLE FROM SPACE

+ + +

In a word, it isn't. Not from the International Space Station (ISS) anyway, where astronauts have tried to see it without success. However, it does show up—just barely—in photos taken from the ISS with 180mm and 400mm telephoto lenses, equipment similar to what a photojournalist would use to shoot a baseball or hockey game.

In November 2004, Chinese-American astronaut Leroy Chiao, determined to put the matter to rest, armed himself with a camera and long lenses and captured the first definitive photos of the wall. His images show segments of the structure in Inner Mongolia about 200 miles (320 km) north of Beijing. One reason the Wall has proved so difficult to see is because it's constructed of materials similar in color to the surrounding landscape and it blends right in. Other smaller man-made features such as desert roads in Australia *are* visible without optical aid from orbit because their color strongly contrasts with the surroundings.

According to NASA, Chiao's photos were "greeted with relief and rejoicing" by the Chinese. As for Chiao, he never saw the Wall and wasn't positive it even showed up in his photos. Others with the agency who examined it closely confirmed it as the real deal.

The Great Wall is really a series of walls that extends for some 1,500 miles (2,414 km) and ranges in width from 13 to 15 feet (4 to 5 m). The parts that still stand were built during the Ming Dynasty from 1368 to 1644 to defend China against invasions by northern nomadic tribes. The idea that you could see the Wall from space goes back to at least 1904. In a book titled *The People and Politics of the Far East*, Henry Norman writes: "Besides its age it enjoys the reputation of being the only work of human hands on the globe visible from the moon."

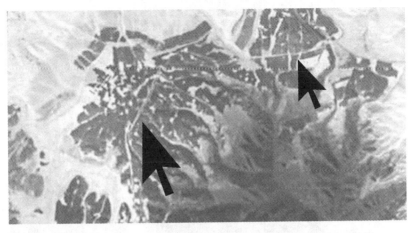

Sections of China's Great Wall are seen in this telephoto image taken by astronaut Leroy Chiao from the International Space Station as it passed over Inner Mongolia on Nov. 24, 2004. (NASA)

Clearly, Norman couldn't know this because the first spacecraft that might check his claim wouldn't be launched for more than half a century. More likely, he was simply trying to convey the magnitude of the structure and literally reached for the moon! Later, in 1938, American adventurer Richard Halliburton reinforced the idea in his book *Second Book of Marvels—the Orient*, writing: "Astronomers say the Great Wall is the only man-made thing visible to the human eye from the moon."

Wrong again. Fortunately, we've now visited the moon multiple times and know firsthand what you can see of Earth from that distant vantage point. Here are the late astronaut Alan Bean's impressions. Bean was the lunar module pilot on *Apollo 12* and the fourth person to walk on the moon:

> The only thing you can see from the moon is a beautiful sphere, mostly white, some blue and patches of yellow, and every once in a while, some green vegetation. No man-made object is visible on this scale. In fact, when first leaving Earth's orbit and only a few thousand miles away, no man-made object is visible at that point either.

If you're disappointed to learn the Great Wall is invisible both from the moon and from orbit, here's a consolation prize—lots of other human-made projects *are*. From the space station's altitude of 250 miles (400 km), they include the lights of hundreds of cities on the globe's night side, some large highways, dams, mining operations and airport runways.

Apollo 17 *astronaut Jack Schmitt stands by the U.S. flag with the Earth shining in the lunar sky on Dec. 19, 1972. From the moon only clouds and the general outlines of continents are visible with the naked eye. (NASA)*

One of the more curious things astronauts can spy is a vast complex of greenhouses near the city of Almeria in southern Spain. It's the largest concentration of plastic greenhouses in the world, big enough to earn the dubious nickname "Sea of Plastic." The sprawl covers 185 square miles (480 square km) and looks like a white bump jutting out from Spain's southern coast.

Before too long, some of us will be going for rides into space as private enterprises like Virgin Galactic and Blue Origin make space tourism both safe and affordable. In case you're one of those planning to buy a ticket, make sure your flight clears the Kármán line, the legal boundary of outer space defined as 62 miles (100 km) above sea level. This is the altitude at which the air is too thin to support a flying aircraft. It's named for Theodore von Kármán, a Hungarian-American engineer and mathematician who first did the calculations.

Jonathan McDowell, an astrophysicist at the Harvard-Smithsonian Center for Astrophysics, takes issue with that definition. In a 2018 paper based on a study of some 43,000 orbiting satellites, he argues that the boundary is a bit closer, more like 50 miles (80 km) altitude. Curiously, this is very close to the limit Kármán originally calculated, 52 miles (83.8 km) and the lowest altitude at which a satellite can still orbit the Earth without quickly falling back into the atmosphere. Whichever figure is ultimately adopted, there's very little of Earth's atmosphere left at either altitude. Dress appropriately and don't forget to bring binoculars if you hope to see that Wall.

WATER GOES DOWN THE DRAIN DIFFERENTLY IN DIFFERENT HEMISPHERES

+ ✦ +

So, is it true?! No book about urban myths would be complete without addressing this burning question. The answer is a qualified "yes."

Water going down the drain in opposite hemispheres swirls down in random directions affected far more by the shape of the drain and the initial "twist" of the liquid than by the hemisphere and spin of the Earth. These local irregularities acting on a very small scale are much more powerful drivers than the spinning Earth, which affects large systems like hurricanes and air masses. As far as your tub is concerned, we can probably forget about experimental proof, but under idealized circumstances—an absolutely still tub of water in a room with no air currents to ruffle the surface—the liquid should theoretically drain consistently in one direction in one hemisphere and in the opposite direction in the other hemisphere due to the Coriolis effect.

Before we talk Coriolis, you first need to know that the Earth spins at different speeds depending on where you live. At the equator, it rotates about 1,000 mph (1,600 kph); at 40° north and south latitudes, 800 mph (1,300 kph); at ±60° latitude at 500 mph (800 kph) and zero miles an hour at the poles. The reason for the difference in speed is because a cup of coffee on the equator, where the Earth is widest, has to travel around a much bigger circle in 24 hours compared to the cup parked next to your computer at a coffee shop in Anchorage. To accomplish this, it must move faster. At the poles, a cup doesn't have to circle around anything—it just sits in one place!

Now we're ready to delve into the Coriolis effect. Pickle Lake, Ontario, and the Galapagos Islands share the same 90° west longitude but have very different

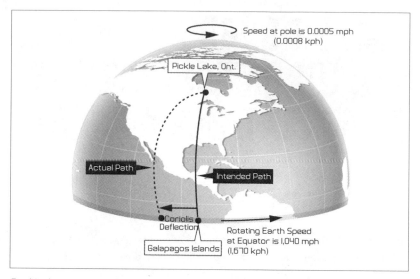

Earth's changing rotation speed with latitude causes an object thrown south from the northern hemisphere to be deflected to the right (west). Objects in the southern hemisphere thrown to the north are deflected to the left (right). Called the Coriolis Effect, it affects which way storms spin and ocean currents flow. (Gary Meader)

latitudes—the Galapagos straddle the equator 3,600 miles (5,800 km) due south of Pickle Lake at 51° north latitude.

Let's say you fire a cannonball from Pickle Lake at a target in the islands. As the cannonball whizzes in the direction of the equator, the speed of the Earth increases slowly beneath it. In fact, the Galapagos spins around the Earth 385 mph (620 kph) faster than Pickle Lake, so that by the time the ball lands, it's no longer due south of where it was fired, but to the right or west of the islands. Next time you fire the cannon, you'll need to account for this and aim east of south to reach your target. Welcome to the Coriolis effect!

Now point your cannon north and fire another cannonball, this time toward the North Pole. It follows the same curving trajectory, but because the ground *slows* beneath the cannonball during its flight, the path twists to the east, or to the right. Similarly, cannonballs fired north or south in the Southern Hemisphere don't land where you'd expect them to either.

A low pressure system over Iceland swirls in a counterclockwise direction due to the Coriolis Effect in this satellite photo. (Aqua-MODIS / NASA)

No forces are at play here. Nothing is acting on the cannonball to curve its path. It's simply the result of the changing speed of Earth's surface. As applied to the atmosphere, the Coriolis effect makes storms swirl counterclockwise in the northern hemisphere and clockwise in the southern hemisphere. Next time a hurricane hits in either hemisphere, take a look at time-lapse satellite photos and you'll see what I mean.

The effect is easy to see in big systems like air masses and ocean currents over long periods of time, but when it comes to your average bathtub, it's too subtle to detect in such a small space over the short amount of time it takes the water to drain. Things like air movement, changing temperature, water currents, surface texture of the tub and whether it's level or not can easily mask the Coriolis effect.

But under experimental conditions, scientists *have* observed the Coriolis effect at work. Ascher Shapiro, a mechanical engineering professor at the Massachusetts Institute of Technology, performed an experiment in 1962 to prove once and for all whether water flowed counterclockwise down a drain in the northern hemisphere.

He set up a circular, flat-bottomed tank with a ⅜-inch (1-cm) drain hole and attached a 20-foot (7-m) hose to the bottom of it. After stoppering the hose, he filled the tank with 6 inches (15 cm) of clean, room-temperature water. Before pulling the plug to let the water drain, he took great care to minimize disturbances like air movement and temperature changes by covering the tank with a sheet of plastic and carefully controlling the room temperature. Finally, he poured the water in with a clockwise swirl, so that if the tank drained counterclockwise it wouldn't have been affected by how the water was added.

He let the tub sit for 24 hours and then pulled the plug while gently placing two crossed slivers of wood above the drain. At first nothing happened, but about 15 minutes into the 20-minute draining process the float began to slowly rotate counterclockwise, reaching a peak rate of one spin every three to four seconds. Repeated trials showed the same result. Eureka!

Shapiro's study was published in the journal *Nature* and was soon verified by other scientists, including Lloyd M. Trefethen, who in 1965 set up a similar experiment in the southern hemisphere and proved that water swirled down the drain in a *clockwise* direction.

I tried my own completely uncontrolled experiment under typical bathroom conditions and used a wooden float to determine the spin. During the first try, the wood pivoted in a counterclockwise circle, which got me very excited. But after repeated trials, the results were mixed.

There you have it. Definitive proof that it does happen under carefully controlled conditions. Just not in *your* bathroom.

METEORS ARE FALLING STARS

The term *falling star* to describe a meteor is so common, it may surprise you to learn that meteors have nothing to do with stars. Except in one way. They look like you'd imagine a star to look were it to detach from the heavens, fall to Earth and burn up in a flash. Good thing that never happens. Stars are massive and bright. Before one could fall to Earth, it would have to be close by, in which case it would shine like a second sun in the sky.

OK, so maybe they don't fall to Earth but across space (the sky) instead. Alpha Centauri, the closest star system after the sun, is so far away that moving at its current speed of around 50,000 mph (80,500 kph) it would take more than 2,300 years to travel the width of a full moon across our sky. That would be one *slow* "meteor."

These days we know that meteors come from much closer to home, namely the realm of the planets. Despite their dramatic appearance, most are very small, similar in size to the little rocks that get stuck in your shoe or the crunchy nuggets in a box of Grape-Nuts cereal. Fainter meteors are no bigger than sand grains. All these dribs and drabs originate from comets, which rubber-band from deep space to Earth's vicinity, and the main belt asteroids, located between Mars and Jupiter.

Asteroid collisions spray rocks everywhere, some of which end up in orbits that intersect Earth's orbit. When the two meet, the nugget slams into the atmosphere and creates a flash of light, or meteor. Comets are primarily made of ice embedded with dust and rocky grit. When a comet passes near the sun, solar heating vaporizes some of the ice, releasing dust and gases that form the comet's fuzzy head and tail. These materials gradually drift away from the comet and spread along its orbit. When Earth slams into the debris during its orbital travels, it salt-and-peppers the upper atmosphere and incandesces as meteors.

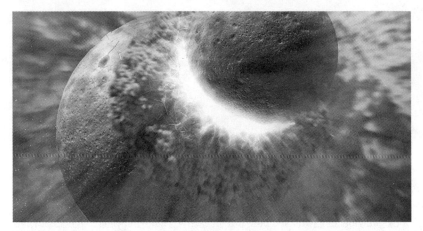

Asteroid collisions send chunks of rock flying, some of which may ultimately find their way to Earth and streak across the sky as meteors. (NASA / JPL-Caltech / T. Pyle [SSC])

Particles and rocks shed from asteroids and comets are called *meteoroids* before they enter the atmosphere and *meteors* when they do. If a meteoroid is large and dense enough to survive its atmospheric pounding and make it to the ground, we call it a *meteorite*.

A meteoroid first enters Earth's atmosphere between 50 and 70 miles (80 and 113 km) high and traveling at an incredible speed, in excess of 45,000 mph (72,000 kph). Friction with the air heats the object to around 3,000°F (1,650°C), quickly vaporizing it into the finest powder called *meteoric dust*. The streak of light we see—the falling-star part—isn't burning per se. The incoming grains heat and compress the surrounding air, exciting air molecules in their path to pop free an electron and become ionized. An ion is an atom or a molecule that has lost an electron and become positively charged. These "ionization trails" are excellent reflectors of radio energy; ham radio operators use them as temporary reflecting "surfaces" to bounce radio signals to distant corners of the globe and converse with old friends and new.

A comet-like trail of debris streams from the object P/2010 A2 in the aftermath of an asteroid collision that took place in 2009. It's estimated that modest-sized asteroids smash into each other about once a year. (NASA, ESA, D. Jewitt [UCLA])

When the molecules retrieve their electrons, they release that pent-up energy as a luminous streak we see as a meteor. There's no actual burning happening in a meteor trail. What we see is a tube of glowing plasma akin to the neon that glows in a beer sign. Each streak lasts about a second and measures 1 yard (1 m) in diameter and several tens of miles long. Scaled down to something you could hold in your hand, it would look like a strand of cooked spaghetti. If you could slow down time a hundredfold and shrink yourself to the size of a mouse, you'd patter through a glowing tunnel of soft light, colored green, yellow or red, depending on the composition and speed of the meteoroid.

Just for fun, let's pretend meteors really are falling stars. How long do you think it would take to empty the sky of all its suns? There are approximately 9,100 stars visible with the naked eye across the entire planet from dark sky sites. Outside of meteor showers like the Perseids or Geminids, about a half dozen random or *sporadic* meteors are visible per hour on a given night to a single observer. Assuming a 9-hour-long night, the sky would be emptied of stars in only 163 nights. Thanks to constant replenishment from comets and asteroids, we needn't lose a single one.

METEORITES ARE BLAZING HOT WHEN THEY HIT THE GROUND

+ ✦ +

People witness and report an average of eight to ten meteorite falls per year. Only rarely is someone close enough to a fallen meteorite to touch it right after it's struck the ground. As we'll learn, reports vary on whether these newly arrived fragments from the asteroid belt feel hot or cold.

One thing we know for sure is that the fiery appearance of bright fireballs that ultimately produce meteorites does not mean they're *on fire*. As we learned in the previous section, their swift passage through the atmosphere excites air molecules to glow for a few brief seconds. In the case of larger meteoroids, the glow can be brilliant, even as bright as the sun based on eyewitness reports. But what look like flames and fire are glowing tubes and columns of ionized air, not flames of combustion like a flaming torch or tree.

It's not easy to shake the Hollywood image of a flaming meteorite or asteroid striking the Earth, but cosmic rocks don't start on fire. First of all, they're rocks. Unlike masses of molten magma hurled from a volcano, they originate in the strange, hot–cold environment of outer space. Here, one side of a meteoroid absorbs the heat of the sun while the night side quickly radiates it away into space. They rip through our atmosphere in a matter of seconds before striking the ground. Friction with the air heats and melts only the outer skin of the meteoroid. The melted material sloughs off in a process called *ablation*, carrying away much of the heat of entry.

Exactly the same thing happens when a manned space capsule returns astronauts to Earth. During reentry the capsule's blunt end experiences temperatures around 5,000°F (2,760°C), similar to that experienced by an incoming meteoroid. How do astronauts survive? Material melts and ablates from the shield, fending off the heat and protecting the precious cargo inside.

This dashcam image shows the Chelyabinsk meteor as it blazed over western Russia at 9:30 a.m. local time on Feb. 15, 2013. At an estimated 66 feet (20 m) across, it was the largest known natural object to enter Earth's atmosphere since 1908. (Aleksandr Ivanov)

Meteorites behave identically. Thickening air at lower altitudes quickly retards the meteoroid. At a height of around 9 to 12 miles (15 to 20 km), the fragments have slowed to 4,500 to 9,000 mph (7,200 to 14,500 kph), the point at which ablation stops and they no longer give off visible light. From there to the ground meteorites are said to be in "dark flight" because they can't be seen. The fragments also cool down quickly because the average air temperature at that altitude is around -60°F (-51°C).

Despite what appears to be a fireball landing over the next hill, a meteoroid still has a minimum of a dozen miles of travel before landing. Most never do—they're either ablated away or fragmented into dust and pieces too tiny for anyone to find. The precious few that reach the ground are coated in a thin, glassy layer of *fusion crust*, which resembles the glaze on ceramic ware. Most fusion crusts are black or dark brown and composed of rock melted during the final seconds of the fall. The fact that meteorite crust is only fingernail-thick (1 to 2 mm) demonstrates how little of the meteoroid is melted during its descent to Earth. Beneath the crust we find the original asteroid rock intact, just as it was before its fateful arrival on a new planet.

Before a meteoroid enters the atmosphere, it orbits the sun in the virtual vacuum of outer space. Just as the airless moon is heated by the sun to hotter than the

One of many Chelyabinsk meteorites that resulted from the Feb. 2013 fireball. It sports blackened "fusion crust" from atmospheric heating and a pale, stony interior. (Svend Buhl / Meteorite Recon CC BY-SA 3.0)

boiling point of water, meteoroids in Earth's vicinity are also warmed by sunshine even in airless space. Studies done that take into account how much radiation the sun pours out at Earth's distance and how the stony and iron meteorite materials absorb that energy show that light-colored, rocky meteoroids have inside temperatures of around 26°F (-3°C). Dark rocky materials are around 98°F (37°C) and irons are the warmest, at 200°F (93°C).

These are approximations. Temperatures vary according to size and surface texture. But assuming that only the outside of a meteoroid is heated during its fall and chilled during the bitter cold of dark flight, then iron meteorites should be rather warm to the touch (but nowhere hot enough to spark a fire), while stony meteorites should range from ever so slightly warm to downright cold. Examples of each have been reported: the Dhurmsala stone that fell in India in 1860 was covered in frost, while the Cabin Creek, Arkansas, iron of 1886 proved "as hot as men could handle."

The Viñales, Cuba, stony meteorite fall of February 1, 2019, is instructive. Michael Farmer, a world-renowned meteorite hunter, shared these reports from the local people he met who picked up meteorites within seconds after they fell:

One boy on a bicycle was riding home rapidly, afraid from the explosions. A large 250-gram meteorite slammed into the asphalt road in front of him, nearly hitting him. He saw it bounce into the ditch and he jumped off the bike and grabbed it, rapidly throwing it down because the stone was painfully cold . . . three other people who saw stones hit the ground and picked them up instantly all said [the] same thing: that they had to drop them because they were so cold.

Given that 95 percent of observed meteorite falls are stony, the majority of space rocks should *not* be hot upon landing, and yet many eyewitnesses describe them as either still warm or too hot to touch. The disconnect between the true temperature of a newly arrived space rock and our perception of it may be connected to our expectations. If you're unfamiliar with the details of falling space rocks and see a fiery fireball breaking apart over your head, you'd expect the pieces to land red-hot. Given that assumption, you might instinctively pull back your hand at first touch, imagining heat.

Expectations or state of mind can sometimes lead to false impressions. Small noises heard in the forest at night can make us think something large and dangerous is moving about when it's nothing more than a field mouse rattling through dry leaves in search of food. Evolution has equipped us with an instinct for potential danger, whether real or imagined.

So few meteorites have been picked up immediately after falling that it's hard to be 100 percent certain one way or another. We can only rule out that they spark fires or come down in flames, except in the case of a massive object which would release enough energy on impact to cause widespread destruction. If you're so lucky as to have a meteorite land safely nearby, pay close attention to what you see, feel and smell and get back to me about your experience.

Resources

- "Temperatures of Meteoroids in Space": http://adsabs.harvard.edu/full/1966Metic...3...59B

WE SEE THE SUN AT SUNRISE

What a silly thought. Of course we see the sun at sunrise. What else would that glaring, orange ball be? What if I told you it was only an optical illusion? We don't ordinarily think of Earth's atmosphere as a prism, but the two have one thing in common. Both bend light. A glass prism bends or refracts the path of a sunbeam twice, first on entry and then on exit. White light is composed of all the colors of the rainbow, with each color refracted to a different degree. Violet gets bent the most and red the least. As sunlight exits the prism, it fans into a classic rainbow according to color.

Air can bend light too. Most of the planet's air is confined to a thin shell only about 10 miles (16 km) thick. When the sun is near the horizon, it shines through the bottom or densest part of the atmosphere. The denser the air, the stronger its refractive power. The reason the sun and moon look oval instead of circular when they rise is because air closer to the horizon "lifts" the more strongly refracted bottom half of the disk upward into the less refracted top, squishing it into an oval. Both morph from circle to sausage as they approach the horizon at sunset. At sunrise, the reverse happens as refraction lessens with the sun's increasing altitude.

Refraction amounts to only a modest lifting of the sun's lower edge when it stands 5° (three fingers held together at arm's length) above the horizon. But by the time the bottom edge of the sun *touches* the horizon, refraction is so strong that it lifts the entire sun into view after it has technically set. You're reading that correctly—there is no sun present within a minute or two of sunset and sunrise. If we could wipe away the atmosphere just as the sun was creeping above the eastern horizon at sunrise, it would disappear for a minute or two until Earth's rotation brought it back into view.

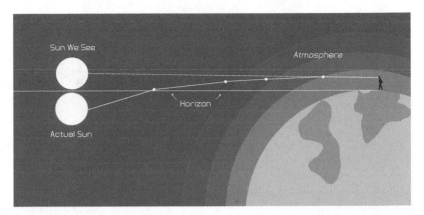

Air can bend light just like a prism does, "lifting" or refracting the sun into view about two minutes before it actually rises, or keeping it in view for the same amount of time when it's setting. (Bob King)

Atmospheric refraction keeps the sun in view longer at sunset and pushes it up before sunrise, tacking some four additional minutes of daylight into each day: two in the evening and two in the morning. Sunrise and sunset times published in newspapers, almanacs and websites already take refraction into account, so the times shown are when the first bead of sun crests the horizon.

I don't get to see many sunsets where I live because of trees, but there's a big lake nearby that tempts me with a new sunrise every clear morning. I'll sometimes make the short drive to watch the first beams crest the eastern horizon. Each time, I get a secret thrill knowing that those first golden rays are completely fictitious, a magic trick performed by nothing more than air.

Refraction affects all celestial objects, lifting the stars and planets at the eastern horizon into view sooner and holding on to them a bit before they drop below the western horizon. On the moon, where the sun rises as a perfect circle because there's no air to warp its shape, you get what you see! Mars has an atmosphere, but it's a hundred times thinner than Earth's, so sunsets and sunrises there suffer little distortion. Planets with thicker atmospheres have more squishing power, enough to compress the sun into a lentil. Strange to think that something as seemingly insubstantial as the air can profoundly affect the appearance and visibility of celestial objects.

EARTH SPINS ONCE EVERY 24 HOURS

You've got to be kidding. How could we have the length of a day wrong? Not that it's wrong exactly, but the 24-hour day we measure is Earth's rotation period with respect to the sun. The length of time between noon one day and noon the next is 24 hours.

Earth's true rotation rate with respect to the fixed background stars is shorter, just 23 hours 56 minutes and 4 seconds, a length of time called the *sidereal day*. Sidereal derives from the Latin word *sidereus*, meaning "heavenly" or "relating to stars." The difference between the sidereal and the solar day length amounts to 3 minutes 56 seconds. That's the extra time the Earth has to rotate on its axis to bring the sun back to the noon position for an earthbound observer.

The reason it takes longer is because the Earth is also *revolving* around the sun as it spins. A spin takes a day, but a revolution takes a whole year. Sensing Earth's daily spin is easy. Just follow the sun's progress from east to west during the day or the stars and moon at night. The planet's yearly revolution is trickier because we can't see the sun move against the background stars, which are invisible by day. But if we could snap our fingers and get rid of the atmosphere, we'd see it slowly crawl eastward (opposite to its daily east-to-west travels caused by Earth's spin) across the background stars, making a complete circle around the sky in a year's time. Of course, the sun only *appears* to be moving—it's really the Earth that moves in its orbit around the sun.

So the only way to keep facing the sun at the end of each sidereal day is to spin a little extra to compensate for that changing perspective caused by the planet's revolution. That amounts to just shy of 4 minutes, which, if added to its true rotation rate, comes to 24 hours. If we decided to live by sidereal instead of solar days, our clocks would soon be out of sync with the sun. It would only take half a revolution (183 days) for twelve noon on our clocks to indicate the sun's position at midnight. Half a revolution later, clock noon would again equal true noon like that dead clock on the wall that's right twice a day.

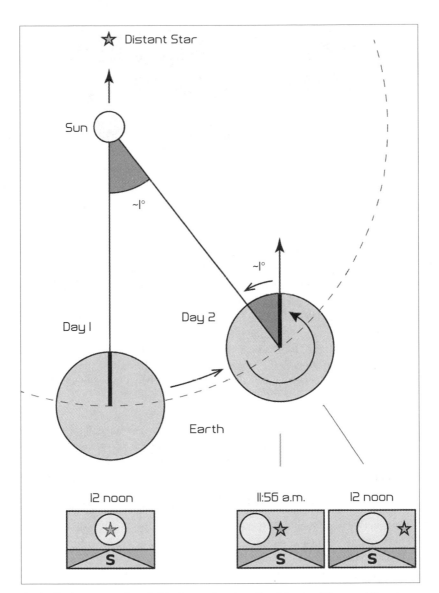

As the Earth rotates on its axis it's also moving along its orbit around the sun at the rate of about one degree a day. With respect to the "fixed" stars, Earth rotates once every 23 hours and 56 minutes, a period called the sidereal day. In the diagram, an observer sees the sun due south at 12 noon on Day 1. One day later, the sun arrives at the same spot at 11:56 a.m.—4 minutes earlier. To bring the sun back to the noon position, Earth has to rotate an additional 4 minutes every day, making the day we use 24 hours long, also known as the solar day. (Xaonon CC BY-SA 4.0)

EARTH'S AXIS FLIPS FROM SUMMER TO WINTER

Most of us have heard that what causes the seasons isn't Earth's distance from the sun. In fact, the planet edges several million miles *closer* to the sun during winter in the northern hemisphere than in summer. The real reason for the change of seasons is the tilt of Earth's axis, equal to 23.5°, about the amount you'd lean forward into a strong wind. On the first day of summer, the north end of the axis along with the northern hemisphere are tipped *toward* the sun, which causes it to appear higher in the sky. The heat striking the ground from a high sun is much more intense. Coupled with the longer days that result from the long, steep arc the sun must travel from sunrise to sunset, many forsake the outdoors for the comforts of central air.

At the same time the northern hemisphere is tipped toward our star, the southern hemisphere is tipped *away* from it. From Down Under, the sun shines low in the northern sky, days are short and sunlight less intense. Winter is underway. Six months later, the situation is reversed. Now the northern hemisphere is tipped away from the sun and in winter's grip, while southerners head for the beach.

Superficially, it sounds as if Earth's axis flips one way and then the other, but we all better hope our planet never does something so drastic. For a flip to occur, a massive asteroid would have to strike the Earth at a certain angle and speed. Astronomers think that's how it arrived at its present tilt in the first place. Instead, Earth simply putters along its orbit with its north axis always pointing in the same direction in space, which happens to be at Polaris, the brightest star in the Little Dipper.

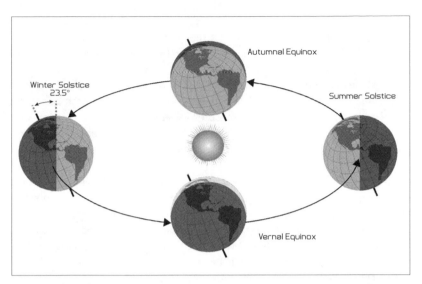

As the Earth revolves around the sun its axis always points in the same direction. Seasons are caused by the changing orientation of the planet with respect to the sun during its yearly orbit. (Gary Meader)

The seasonal change is caused by the orientation of Earth's tilt toward the sun, which changes during its yearly orbit. All Earth has to do is keep going around the sun and the orientation takes care of itself, as you can see from the diagram. On one side of its orbit, the north axis leans toward the sun. Steadfastly pointed toward Polaris, it continues to the other side of its orbit, and look what happens: the pole and northern hemisphere now tilt away from the sun.

The start of winter and summer, called *solstices*, mark the extremes of what our titled axis can muster: 23.5° toward or away from the sun. Midway between those extremes are the *equinoxes*, better known as the first days of spring and fall. At these times, Earth is sideways to the sun, with neither hemisphere tipped toward or away. We're at a balance point. Everywhere on the planet the sun rises in the east and sets in the west. Day and night are of equal length. The sun is neither high nor low in the sky but midway between its winter low and summer high.

As the planet continues on its merry way around the sun, day and night soon fall out of balance. The day after the fall equinox, night begins to overtake day as Earth's northern hemisphere moves inexorably toward winter. After the spring equinox, day length increases and bites into the night.

You might be surprised to learn that Earth's tilt is not fixed but varies between 22.1° and 24.5° over a 41,000-year cycle caused by the gravitational attraction of the sun, moon and planets. Changes in the tilt of the axis affect the intensity of the seasons and may contribute to the waxing and waning of the Ice Ages. Lesser tilts distribute the sun's heat more evenly across the planet, moderating temperatures. Steeper tilts lead to more extreme seasonal variations with hotter summers and colder winters. The tilt is currently decreasing from its last maximum around 9,000 B.C.E. and will reach the next minimum about 11,800 C.E.

If Earth had no tilt but rotated straight up and down, there would be no seasons. We'd experience something like the people who live in the tropics do now—one continuous season of similar temperatures, cloud cover and day length throughout the year. Areas near the equator, where the sun would be overhead at noon every day, would certainly feel the heat. Mid-latitudes would experience perpetual fall-like or spring-like temperatures, while the polar regions would forever be cold but without the bitter extremes we see today. All days would also be nearly equal in length across the planet: 12 hours of daylight and 12 hours of night.

Despite common complaints about seasonal extremes, I think we'd all grow bored of this seasonal monotony. We can thank our lucky asteroids for our tipped axis. Earth, like the other planets, was heavily bombarded by planetesimals, minute planets that collided and came together to build the planets as we know them today. One or more of them slammed into the proto-Earth and knocked it awry. Similar blows presumably affected the other planets, leaving some spinning nearly straight up and down, like Mercury and Jupiter, while Mars and Neptune have tilts quite similar to that of Earth. Uranus spins sideways!

Tilted axes don't seem like a big deal, but the happenstance that left Earth listing has led to the incredible diversity of life across radically different climate environments.

GALILEO INVENTED THE TELESCOPE

✦ ✦ ✦

Italian astronomer Galileo Galilei did so much to advance our understanding of the planets, moon, sun and Milky Way with his handcrafted telescopes, it's natural to think he invented the instrument. He didn't. He just built really good ones and thought of amazing ways to use them. His first instrument was modeled on the three-power telescopes being sold in Paris and other locales in Europe in the year 1609.

The real inventor may have been the Dutch spectacle maker Hans Lippershey, who allegedly placed a convex lens at one end of a tube and either a convex or concave lens at the other end to make an instrument that magnified three times. Lippershey applied for a patent for his Dutch *perspective glass* "for seeing things far away as if they were nearby" to the States General of the Netherlands on October 2, 1608. It was never granted because of competing claims by two other Dutch eyeglass makers, Jacob Metius and Zacharias Janssen. Officials also noted that the design was so simple and easy to make, it would be difficult to patent. Lippershey received a sizable sum of money from the government for copies of his design, while the *telescope*—a word that Greek mathematician Giovanni Demisiani coined in 1611—was set free into the world.

Galileo, an Italian physicist and mathematician, first got wind of the new technology in late July of 1609 and quickly fashioned one of his own. He claimed that he teased out the design based on the "laws of refraction," but it seems more likely that he corresponded and met with others who explained the design in some detail.

Galileo was a master at grinding and polishing lenses and quickly refined his telescopes to the state of the art for the time. His next instrument magnified 8x (some sources give 9x), then 20x and finally one that reached the unheard-of magnification of 30x.

1754 painting showing Galileo describing his telescope to Leonardo Donato, chief magistrate of Venice, and the Venetian Senate. (H.J. Detouche)

His handmade telescopes were works of art, but the images they provided were poor by today's standards. They suffered from optical defects called *aberrations* that distorted and added false color to the images. Their fields of view were so tiny that he might as well have been looking through a soda straw. When we see Galileo's wonderful drawings of the moon, we can appreciate how he had to "stitch together" multiple views to record his final impressions of the complete lunar disk.

On August 24, 1609, less than a month after he built his first telescope, Galileo presented an 8x version to the senate in Venice, Italy. He smartly pointed out that users could identify an enemy at sea much sooner with a telescope than by eye alone. The Senate must have been impressed because they rewarded him with tenure and doubled his salary as professor of mathematics at the University of Padua.

Galileo was a great self-promoter and public relations genius who claimed he intuited the telescope design (debatable) and built it to high specifications (true). Because it was the first working telescope the senators had laid their eyes on, from their perspective, Galileo *was* the inventor. This and Galileo's insistence that he cooked up the design on his own after hearing about it still muddle the true story of its origin. Add in the professor's long list of astronomical firsts that forever changed how we see our place in the universe, and it's not surprising that some still mistakenly see him as its inventor.

A reproduction of one of Galileo's telescopes. (Leonardo da Vinci National Museum of Science and Technology, Milan)

With his great skill as a telescope maker, Galileo made money on the side selling his spyglasses to merchants and others in Venice, while taking his own instrument and pointing it at stars instead of ships. Galileo was the first to observe the four bright satellites of Jupiter, still called "the Galilean moons" to this day, and to discover the phases of Venus. He also studied Saturn and sunspots, and revealed the starry nature of the Milky Way.

For centuries, people thought the moon was a perfect sphere in a distant realm. Galileo must have been shocked at his first look at the moon in November 1609. Instead of a smooth, crystalline sphere, he gazed upon a rugged landscape wrinkled by mountains and creased by valleys that more resembled the Earth. Heavenly bodies were supposed to be perfect and unchanging, but clearly the moon was neither. And if the moon wasn't, then maybe the planets weren't either. Practically overnight, the perfect heavens became imperfect. Thanks to the telescope, people came to realize that the moon and planets were our relatives, not parts in a perfect machine. What a shift in consciousness this little brass tube inspired!

You may have heard that Galileo was the first person to study the sky with a telescope. He was not. Recent research by historian Allan Chapman of the University of Oxford has revealed that the English astronomer and mathematician Thomas Harriot peered through his new 6-power "Dutch truncke" (telescope)

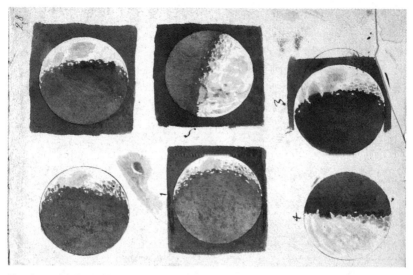

Sketches made by Galileo with his handmade telescope in the fall of 1609. (Public domain)

at the five-day-old moon and made a sketch of it dated July 26, 1609—several months before Galileo took to looking at the sky. Harriot never published his drawings or pressed his claim, but the fact stands for the record.

Galileo neither invented the telescope nor was he the first to point it at the stars. But he did have a scientific mind that led him to record what he saw, put forth new hypotheses and share his discoveries through books, demonstrations and talks. His careful observations led to a sea of change in our sense of scale and place in this cavernous cosmos we call home.

Resources

- Galileo timeline: http://galileo.rice.edu/chron/galileo.html

- "How did Galileo develop his telescope?": https://innovation.ucdavis.edu/people/publications/biagioli-did-galileo-copy-the-telescope

- "Galileo and the Telescope": https://www.loc.gov then search "Galileo and Telescope"

OBJECTS IN A TELESCOPE LOOK LIKE THEY DO IN PHOTOS

+ ✦ +

Do those beautiful photos that light up websites and fill the glossy pages of astronomy books whet your desire to purchase a telescope? They certainly nudged me toward the hobby. That expectation leads some telescope buyers to expect similar mouthwatering views when they step up to look through the lens. Let me advise caution. Cameras reveal eye-watering details and colors in nebulas and star clusters for one good reason—they can accumulate the light from otherwise faint, dull objects during *time exposures* and transform them into bright and colorful images. Our eyes can't do this—we see things in *real time*. A faint object looks faint and no amount of staring will make it appear brighter.

There are some little tricks we can play to make a dim sky object appear momentarily brighter. Our retinas have two types of light-sensing cells: cones for color vision and rods for black-and-white night vision. Cones are concentrated in the center of the retina and used for daytime vision. Rods can't sense color and aren't much good at fine detail either, but they excel at seeing in the dark and sensing motion.

To make the best use of the rods, which are more densely concentrated a short distance from the center of the retina, skywatchers use their peripheral vision, better known in the hobby as *averted* vision. Instead of looking directly at the object, you look "around" it, letting the light fall where the rods are most concentrated. Instantly, the object brightens up into better visibility. To use averted vision most effectively at the telescope, look slightly rightward when using your right eye and leftward with your left eye. It also helps to roll your eyeball around to refresh the view.

One of the Hubble Space Telescope's most iconic images—the dusty fingers of the Eagle Nebula better known as the "Pillars of Creation." The pillars are 5 light years tall and composed of dust and gas. Deep within each tendril, new stars are being born. (NASA, ESA, and the Hubble Heritage Team [STScI/AURA])

Outside of planets, which appear colorful and bright but much smaller than depicted in photos, most everything else in a telescope looks like gray fuzz, at least at first glance. That goes for comets, galaxies and nebulae. But on closer inspection and with experience, more details emerge until each object reveals its own unique beauty and character.

Delicate spiral arms emerge from galaxies. Nebulae assume the appearance of flowers, cats' eyes and even horses. A few even show hints of color—subtle blues, greens and reds—and many have stars tucked here and there that lend a wonderful sparkle to the scene. While cameras will still capture far more detail, many images lack the dynamic range of the human eye. Telescope observers can see details in both the bright and the faint parts of a nebula at the same time, while a camera will often overexpose and "blow out" bright details to better show faint structures. Biology has its advantages.

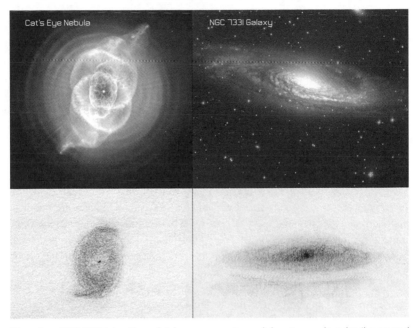

The galaxy NGC 7331 taken through telescopes are crisp and show tremendous detail compared to views of the same objects in an amateur telescope. The sketches were made using a 15-inch (38-cm) (largish) reflector. Visually, objects are much fainter, and except for the planets and moon, look fuzzy. That doesn't make them any less interesting however! (Top left clockwise: NASA, ESA, HEIC and the Hubble Heritage Team [STScI/AURA]; NASA, JPL-Caltech, M. Regan [STScI], SINGS Team / Daniel Bramich [ING] and Nik Szymanek; Bob King)

Views through a telescope have a delicacy and subtlety missing from photographs. To see such things takes time and practice. Just remember that the most important part of observing is that you're seeing the *real thing*, not a reproduction on a computer screen or page. Glimpsing a galaxy that's millions of light-years away, even if it's no more than a faint smudge, is a direct connection between the light of its billions of suns and you. Your retina compiles an image and sends it to the brain, where it's converted into the electricity of wonder. That's my kind of miracle.

I should point out that not *everything* outside of planets looks fuzzy. Open star clusters like the Pleiades (Seven Sisters) pack dozens to hundreds of stellar points together into attractive groups that delight the eye. Globular clusters, dense spherical agglomerations of hundreds of thousands of individual stars, look misty at first sight but, when magnified and carefully focused, overflow with stars like sugar spilled on a black tablecloth.

And then there's the moon. It's the single nighttime object that looks equal to or better than what you see in photographs. I love the moon. As far as I'm concerned, it's worthy of a lifetime of telescopic observation. The moon stands out from everything else because it's so close. You don't need a big scope or high magnification to be absolutely blown away by the views. You'll never run out of craters, cracks, crumbling mountains and lava plains to study.

Small telescopes with mirrors or objective lenses measuring 2.5 to 4 inches (60 to 100 mm) across do well on planets, the moon and brighter star clusters but less so on nebulae and galaxies. Most of those are relatively faint and need the extra light-gathering power of a 6- to 10-inch (150- to 250-mm) scope. Large telescopes excel at revealing structure and elusive colors in the brighter nebulae, spiral arms in galaxies and faint, remote objects like quasars—giant black holes munching on stars and gas in distant galaxies—but they're expensive and often heavy. Anything you have to lug around probably won't see as much use as a smaller instrument.

No matter what telescope you might consider buying someday, don't expect Technicolor views. Do expect a lifetime of exploration. Observing is what you bring to it. Knowing a little about what you're seeing can make the difference between a superficial experience and a deeper one. As my older daughter Katherine once said, "It's all gray fuzz, but it's crazy to think what's out there."

THERE'S NO GRAVITY IN SPACE

+ ✦ +

Appearances can deceive, as can fuzzy language. The International Space Station (ISS) has been in Earth orbit for more than 20 years. You've probably seen crew members on TV floating around the cabin, taking photos while hovering upside down or gulping shimmering globs of water suspended in midair. Everything inside the space station is weightless, and will float away if not tethered.

For those of us who will never get off the Earth, the concept of gravity is a weighty one. Just getting up from a chair we have to use our muscles to counteract the force of gravity, which would happily see us stay put. The heavier you are, the more effort it takes to move. More massive planets make the task even tougher. If you weigh 150 pounds (68 kg) on Earth, on Jupiter you'd weigh 351 pounds (159 kg). Imagine how hard it would be to get off the couch now.

Gravity implies weight, so zero gravity would seem to imply weightlessness, but the two are very different things. When you step on a scale to see how many pounds you've lost or gained, the scale isn't really measuring your weight but rather the upward force applied by the scale against your feet to balance the downward force (pull) of Earth's gravity. A scale is calibrated to spin up a number we call *weight*.

Weightlessness occurs when there's no force of support on your body. With nothing "in the way" to stop you from falling, you'll just keep on plummeting in *free fall*. On Earth we feel the ground or floor pushing up, which prevents us from free falling to the center of the Earth under the pull of gravity. For a taste of free fall, jump off a porch. Before the ground stops your progress you'll be in free fall.

Astronauts, along with their ship, its windows, floors, food packs, beds and everything else, are in continuous free fall. No external force acts to create the sensation of weight. They feel only the force of gravity exerted by the Earth, which acts to pull them straight down toward the ground 250 miles (400 km) below.

Flight Engineer Cady Coleman plays the flute while weightless on the International Space Station. (NASA)

So why doesn't the space station crash into the Earth? Because the ship moves forward in orbit fast enough to defeat the downward pull of gravity. Get enough speed going to compensate for Earth's gravity, and you'll circle almost forever around the curved globe of Earth without fear of hitting the ground. The ISS maintains its orbit by traveling at about 17,150 mph (27,600 kph), more than 22 times the speed of sound. Because there's almost no atmosphere at its high altitude, there's minimal friction or drag on the spacecraft. Only occasionally do the astronauts need to fire the space station's thrusters to "refresh" its orbit; otherwise, it just goes and goes. If it were to suddenly come to a standstill (which it won't), the whole works would plummet straight to the ground. Forward speed makes all the difference, and it's why rockets burn a lot of fuel to launch a probe into a safe and stable orbit.

Because the space station orbits 250 miles (400 km) farther from the planet than you and I do, gravity there is a tad weaker. Although the astronauts are weightless as they orbit, if they could stop the ship, hit pause and slip scales under their feet, they'd discover that they weigh less than on the ground: 11 percent less, to be exact. Can you guess why? It's because they're farther from Earth than you and I and feel its attraction less. The farther space travelers venture from the planet— say, on a moon mission—the less they would weigh, but all would continue to feel the pull of Earth's gravity, however slight, even millions of miles away.

Stephen Hawking, former theoretical physicist and cosmologist, got to experience weightlessness while on board a Zero Gravity Corp. plane in 2007. (Jim Campbell, Aero-News Network)

Gravity may not be a strong force, but it's pervasive. If you double your distance from a moon or planet, its attractive force becomes a quarter as strong. As distance increases, gravity's pull becomes less and less but never goes to zero. Even on the other side of the galaxy, Earth's gravity could theoretically be measured, assuming you have an incredibly sensitive instrument. Every object in space feels the gravitational pull of other objects, but if you're very far away from them, such as in outer space beyond the moon and Earth, then you're virtually weightless because the forces pulling on you are so slight. For example, the sun is 333,000 times more massive than Earth, but because it's so much farther away its gravitational pull is just 0.0006 times that of our home planet.

Most of us have experienced a touch of weightlessness on an elevator. When the elevator rises quickly, the floor pushes up and we feel heavier. When it quickly drops to the floor below, we feel a touch lighter. If the cable were to snap, both the elevator and its occupants would drop in free fall under the force of Earth's gravity, accelerating at the rate of 35.2 feet (9.8 m) per second. Essentially, that's what's happening at the space station—everything and everyone is in free fall.

NASA trains astronauts for life in space by recreating weightless conditions in the famous "vomit comet," a plane that climbs to high altitude and then dives back toward the ground to simulate weightlessness. The agency used to fly Boeing KC-135 Stratotankers for these missions, but they now contract with Zero Gravity

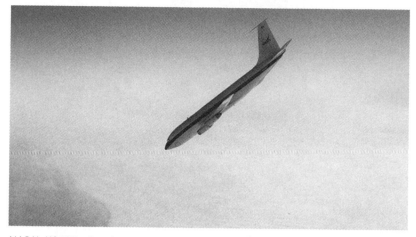

NASA's KC-135, a.k.a. the Vomit Comet, performs a steep dive to simulate weightlessness for its occupants. (NASA)

Corp. for flight on a modified Boeing 727. If you'd like to try it yourself, a seat will set you back $4,950 plus 5 percent tax (as of early 2019) for 20 to 30 seconds of weightlessness.

There are easier and less expensive ways to experience free fall. If jumping off a porch doesn't satisfy, parachute jumping will surely do the trick. Before you pull the cord to release the chute, take in the exhilaration of having little between you and the claws of the planet's gravitational field.

Like the space station, the moon free falls toward the Earth. And the Earth falls toward the sun, but they don't crash into each other because each is moving fast enough to stay in orbit—just like the space station.

Earth is just as weightless in relation to the sun as an ISS astronaut is in relation to the Earth. Each has mass, so each is subject to gravity, especially the gravity of massive nearby objects. Because the sun comprises 99 percent of the mass of the solar system, it has no problem "holding on" to eight planets, countless asteroids and comets. You and I have mass, too: the reason we're willing captives on this watery blue globe.

NASA SPENT BILLIONS ON "SPACE PENS"

+ + +

According to a widely circulated urban myth, NASA invested $12 billion dollars of taxpayer money many years ago to build a pen that could write in weightless conditions, while the then–Soviet Union opted for the simple elegance and utility of the pencil. Pretty dumb of NASA, right? While we can all point to instances of the government overpaying for something when a less expensive alternative was available, that's not what really happened.

In the early years, NASA astronauts used pencils. During Project Gemini, active from 1961 to 1966, NASA purchased 34 mechanical pencils from Tycam Engineering Manufacturing, Inc. in 1965 at $128.89 per unit. When the news went public, people were outraged. In 2019 dollars, that's like paying over a thousand bucks per pencil. NASA got an earful, backtracked and purchased less expensive items.

Pencils had been problematic from the start. Bits of wood would flake off, while tips would occasionally break and float around the cabin, getting into an astronaut's eyes or nose or shorting out electrical devices. Pencils, erasers and the graphite used in the lead were also flammable. Worse yet, graphite conducts electricity, making it an electrical hazard in a space capsule's oxygen-rich environment.

The Soviets used grease pencils on plastic slates as a substitute for wood pencils, but the result wasn't as durable as pen on paper. Regular pens, which rely on gravity to get the ink flowing, weren't of much use under weightless conditions either. What to do?

Paul C. Fisher of the Fisher Pen Co. had the solution: invent a space pen. The company invested a million dollars of its own money—no NASA funding

The Fisher Space Pen model AG-7, which sold for $6 apiece in 1968, was used by NASA and Russian astronauts. (CPG 100 / Wikipedia / CC BY-SA 3.0)

involved—to design and patent a ballpoint pen in 1965 that would operate in space. Named the AG-7 "Anti-Gravity" pen, the device had a pressurized ink cartridge injected with nitrogen gas that functioned in weightless environments and across a wide temperature range from -50°F to +400°F (-45°C to +205°C). The pens used a special ink with a semisolid, gel-like consistency that became fluid as the ball rolled along the paper's surface.

Fisher approached NASA with his new pen, but the earlier controversy over the pricey pencils made the agency a little sheepish, so they held off buying the new device until 1967. After rigorous testing and several modifications, such as wrapping the pens in Velcro to stick to space suit coveralls and walls, NASA included them on the Apollo missions. According to an Associated Press (AP) report from February 1968, the agency bought some 400 pens from Fisher at a reasonable $6 each ($45 in 2019 dollars). Quite a steal.

The Soviet Union also got in on the deal and bought 100 of the Fisher pens plus 1,000 extra ink cartridges in February 1969 to use on its Soyuz space flights in place of the old grease pencils. Both Russians and Americans continue to use Fisher Space Pens to this day on the International Space Station, though some ISS residents like to mix it up a little. NASA astronaut Clayton C. Anderson liked red Sharpies and

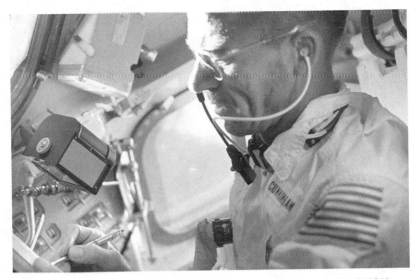

Apollo 7 astronaut Walter Cunningham uses a Fisher Space Pen on a mission. (NASA)

mechanical pencils. Despite the occasional broken lead, the station's ventilation system is so good at cleaning the air that he never ran into a problem.

The greatest role a pen ever played in the history of space exploration happened during the Apollo 11 mission, the first to land men on the moon. Buzz Aldrin, the second person to set foot on lunar soil after Neil Armstrong, writes in his book *Magnificent Desolation* that were it not for a felt-tipped pen, he and Neil might not have made it off the moon and back to Earth.

As they prepared to leave the moon, Aldrin noticed that a circuit breaker switch that activated the lander's ascent stage had broken off. Without the crucial switch there would be no leaving the moon. Aldrin relayed the bad news to Houston, but the next morning had a revelation. He reached for the black, felt-tipped marker in his shoulder pocket and inserted it into the small opening where the switch would have been. After a firm push, the circuit was live, and the crew was able to lift off. The rest is history. Aldrin still has the pen—and the breaker switch, too!

Resources
•"The Saga of Writing in Space": https://airandspace.si.edu/stories/ editorial/saga-writing-space

MOON

WE NEVER LANDED ON THE MOON

✦ ✦ ✦

When I was fifteen years old, caught up in the excitement of the *Apollo 11* moon landing, I could never have imagined that 50 years later people would insist it never happened. Granted, it's only a minority, but enough to be disturbing. Neil Armstrong put his boots down into the lunar regolith—the scientific term for moon dirt—at 10:56 p.m. (Eastern Time) on July 20, 1969. I watched it unfold in real time on a black-and-white TV set in the basement of my home in suburban Chicago. To make sure those first moments wouldn't be lost to history, I set up a tripod in front of the television, framed the scene and captured the gritty black-and-white scenes on a roll of Tri-X film with my Argus C-3 camera.

I *lived* the space program as a boy, and like a lot of kids, I wanted to be an astronaut. While I waited my turn, I regularly mailed letters to NASA asking for free literature about outer space. Brown-paper envelopes with official-looking government return addresses would arrive filled with brochures and photos of spacecraft and the astronauts who would pilot them. My favorite reading haunt was a backyard maple tree. I'd climb and find my happy place between the branches, a little closer to the sky.

The Apollo missions took place in the context of the Space Race, a competition between the United States and the Soviet Union for dominance in outer space that began in the mid-1950s. After slow starts and numerous rocket failures, lots of Americans thought we were falling behind in that competition, so President Kennedy looked for an achievement that could restore America's superiority in space. He chose a daring goal, one that appealed to the country's pioneering spirit. On May 25, 1961, Kennedy stood before Congress and proposed that the United States "should commit itself to achieving the goal, before this decade is out, of landing a man on the moon and returning him safely to the Earth."

People heard the call and got to work. At its peak the Apollo program employed 400,000 Americans, including rocket designers and builders, astronaut trainers,

Apollo 11 *astronaut Buzz Aldrin sets up the Passive Seismic Experiment at Tranquility Base in July 1969. It was the first seismometer on the moon and used to detect moonquakes. Solar panels power the instrument. (NASA)*

science experts and office personnel. In spite of setbacks and snafus, this gargantuan enterprise ultimately met its deadlines and achieved Kennedy's dream.

Newspapers and TV reported on every aspect of the seven missions, including the near-disaster of *Apollo 13*, which suffered an explosion that forced the crew to cut the mission short. When the astronauts returned to Earth, journalists covered every crew and capsule retrieval from the ocean. Media stationed on the ships that would transport the astronauts home captured their arrival by helicopter in pictures, stories and footage that graced magazines, newspapers and television screens across the planet.

Sadly, the president didn't live to see his goal realized, but many of us did, and we'll never forget its significance or the pride and privilege we felt to witness it. How long have we dreamed of going to the moon? And how did I get so lucky to be around to see that dream realized? We did it! We went to the moon! It felt as if the whole world paused to take a breath from strife and relish a shared achievement.

A dozen astronauts walked or hopped (easier in the moon's low gravity) on its surface, took thousands of images, gathered and meticulously packaged 842 pounds (382 kg) of moon rocks, erected flags, tore around in a battery-powered "moon buggy" and even set up a telescope to photograph stars in ultraviolet light. NASA worked the men hard, but there were lighter moments.

Apollo 14's Alan Shepard famously fashioned a makeshift six-iron and sent two golf balls sailing during his lunar stay.

After six missions between July 1969 and December 1972, the Apollo program was terminated. One reason was cost. It took a lot of money to send people to the moon, and the country had met Kennedy's goal of a signature Space Age achievement. Why keep proving the obvious? Don't say that to a scientist. Despite protests from scientific quarters that we had barely scraped the lunar regolith, NASA scrapped the program and invested in other priorities like Skylab, the first orbiting space station.

Starting in the mid-1970s, a few individuals and groups began accusing NASA of faking the moon landings, claiming inconsistencies in some of the photographs, the prohibitive cost of sending astronauts to the moon and the misplaced belief that radiation in Earth's Van Allen radiation belts would kill any astronauts attempting to leave Earth's orbit. Some claimed NASA filmed it all in a movie studio on Earth using clever lighting and props.

Beliefs like these often stem from a general distrust of government-run organizations, or worse, the need to denigrate the achievements of others to enhance one's own self-importance. But how do you fake something that involves more than 400,000 people working across 20,000 industries, universities and government for ten years? Something that was thoroughly documented by worldwide media? Even the "evil enemy," the Soviet Union, conceded the landings happened. They could have yelled "fake!" but didn't because they knew the facts were plain.

Lack of trust may have been the match that set the fire, but misunderstanding of scientific principles and willful ignorance of the evidence have kept the controversy alive for more than 50 years. According to Wikipedia, opinion polls have shown that between 6 percent and 20 percent of Americans don't believe the moon landings happened. You may not be among the disbelievers but still have questions after reading misinformation purveyed online. Or maybe you're just curious to learn more. You've come to the right place.

To judge from some posts I've read, it seems that nothing short of standing on the moon in person would convince Apollo doubters we went there. Consider this. No one today was alive during the Civil War, yet we're confident it took place because it's been thoroughly documented, and we have a mountain's worth of physical evidence stored away in museums and private collections to prove it. Newspapers of the time covered the war the same way newspapers covered the ups and downs

The S-1C booster for the Apollo 11 *Saturn V waits inside the Vehicle Assembly Building at NASA's Kennedy Space Center in February 1969. (NASA)*

of the Apollo program. But unlike the Civil War, the vast majority of us can't take a rocket to the moon to see the landers, experiments and footprints left by the astronauts like we can the monuments, burial places and implements of war. That doesn't make them any less real, just *less accessible*. If someone doubts the existence of things simply because they can't see them in person, they better get ready to deny much of history, including the spacecraft sent to seven different planets plus Pluto.

If you write off Apollo, why not everything else you can't absolutely verify by going there yourself? While you and I will never stand face to face with the Mars *Curiosity* rover or the boot prints left by the Apollo astronauts, the photographs and data gathered by astronauts and our robotic emissaries are available almost 24/7 to see and study online or at a public library.

In one sense, Apollo is even more real compared to other historical events because many of the people who built the program, including a few of the astronauts who walked on the moon, are still around. Journalists can email or dial up an eyewitness for an interview the same way soldiers who survived the Civil War lived well into the twentieth century to share their tales of an event long past.

Back in 2000, I had the pleasure of meeting astronaut Harrison "Jack" Schmitt along with fellow astronaut Gene Cernan, who both spent 75 hours on the moon during the final Apollo mission. I asked what he thought now years later when he looks up at the moon. He flashed a smile and said: "It still catches my eye."

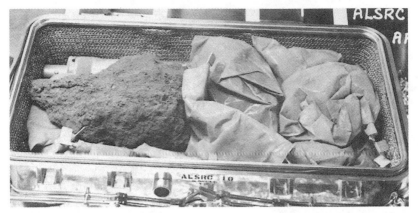

A peek inside the second Apollo Lunar Sample Return Container (ALSRC) shows an assortment of rocks collected by the Apollo 11 *astronauts. (NASA)*

There are many more proofs we went to the moon, so I prepared a list for you. Keep it handy when the moon-landing deniers show up at the door.

10 Definitive Proofs We Really Went to the Moon

I. Rocks, baby! We got the rocks. The Apollo missions returned 842 pounds (382 kg) of lunar rocks, pebbles, dust and core samples for a total of 2,200 samples from six different landing sites. The main repository for all this material is the Lunar Sample Laboratory Facility at the Johnson Space Center in Houston, Texas. In addition, three robotic Soviet lunar missions returned about ¾ pound (300 g) from three other lunar sites. Nearly 400 samples are distributed each year to scientists around the globe for research projects.

    Want to borrow a sample? If you work at a school, go to curator.jsc.nasa.gov/education/public_display.cfm and request samples for educational use. Others can request pieces for public display in a museum or planetarium and for special events. Monster truck show, maybe?

2. Surface maps of the moon made by current satellites using laser altimetry exactly match photographs taken of those same areas by the lunar astronauts. NASA's *Lunar Reconnaissance Orbiter* breaks a single pulse of laser light into five separate beams aimed at the surface. Information in the light scattered back to the orbiter along with the amount of time it's delayed depend upon the roughness of the surface and its elevation. Together, the

On the left is a photo of the landing site at Hadley Rill taken on Aug. 2, 1971 by the Apollo 15 astronauts. At right is a reconstruction of the same site from images and altitude data taken by Japan's lunar orbiter Kaguya in 2008. The two are virtually identical. Kaguya's camera could only see objects down to 33 feet (10 m) wide, the reason the image lacks the detail of the Apollo photo. (NASA [left], JAXA)

measurements provide precise three-dimensional surface elevations across the lunar globe down to about 3 feet (1 m).

3. Japan's *Kaguya* orbiter, which orbited the moon from 2007 to 2009, used a similar setup combined with photography to generate strikingly detailed virtual moonscapes. Using only *Kaguya* data, Japanese Space Agency engineers created a digitally rendered view of the moon as if an observer stood at the *Apollo 15* landing site. It exactly matches the photo above taken by astronaut Jim Irwin in July 1971.

4. If the Apollo landings were faked, the Russians would have pounced on a golden opportunity to discredit the United States. They didn't, and it never came up after the fact either.

5. The Jodrell Bank Observatory team in the UK used the 50-foot (15-m) radio telescope to monitor radio signals from the *Apollo 11 Eagle* lander as it descended to the moon's surface. The readout on graph paper actually shows the moment Neil Armstrong took manual control of the lander, a quick-thinking move to find a smoother landing spot. The team also captured the moment the lander touched down. You can see it in hard black-and-white at the site listed under Resources.

6. Astronauts with the Apollo 11, 14 and 15 missions left retroreflectors called corner-cube prisms on the lunar surface. The *Apollo 11* array consists of 100 prisms on a 2-foot (61-cm)-wide panel aimed at the Earth. Astronomers on Earth ping the arrays with pulses of laser light, then measure the time

Apollo 14's *Lunar Ranging Retro Reflector, placed on the moon's surface on January 31, 1971, has continued to return data from the moon as of this writing.* (NASA)

it takes for the reflected light to arrive back at the telescope, yielding the distance to the moon accurate to within a couple of inches. They've been shooting lasers at them for decades. The 27.5-inch (0.7-m) telescope at McDonald Observatory routinely pings all three arrays to this day—*Apollo 11* in the Sea of Tranquility; *Apollo 14* in the Fra Mauro Highlands and *Apollo 15* at Hadley Rill. Thanks to the reflectors, we've learned that the moon is slowly spiraling away from Earth at the rate of 1.5 inches (3.8 cm) a year due to tidal interactions with the Earth, and likely has a semi-molten core.

7. Photographs taken from the moon's surface show no stars in the sky. Most people would expect lots of stars to show in the pitch-black lunar sky. Without an atmosphere to dim their light, stars should be even brighter than on Earth, right? While the stars do appear somewhat brighter, the astronauts took all their pictures in sunshine using camera settings similar to those they would use on Earth on a sunny day. Given their nearly identical distances from the sun, both bodies receive the same amount of light.

Typical camera settings in full sunshine are very brief, only a tiny fraction of a second, impossibly short to record stars, which require a minimum of several seconds or longer, especially using the slower-speed films of the time. Long exposures would not only have required a tripod but would have totally overexposed (blown out) the sunlit landscape.

8. Dust flying off the wheels of the lunar rovers falls back to the ground differently than on Earth because the moon has no atmosphere and only ⅙th the gravity of our planet. Give a physicist a video of the astronauts doing wheelies and she'd deduce it was happening on an airless body with a fraction of Earth's gravity.

9. We have detailed photos of each Apollo landing site taken by NASA's *Lunar Reconnaissance Orbiter*. Mission engineers can lower the orbiter to about 12 miles (20 km), close enough to resolve footpaths, rovers, descent stages and even a couple of the remaining flags. These remarkably detailed photographs are available for anyone to see online. No, these are not faked. The Orbiter has also photographed the Russian landers as well as the Chinese *Chang'e 4* lander (2019) and rover at the same level of detail.

10. Yes, astronauts did survive the Van Allen radiation belts. You may have read that radiation hazards posed by the Van Allen radiation belts would have killed the astronauts as they attempted to leave Earth's orbit to embark on their journey to the moon. The belts are two doughnut-shaped zones of energetic particles—mostly electrons and protons—trapped from the solar wind by Earth's magnetic field. The inner belt starts far above the atmosphere at around 620 miles (1,000 km) and extends to 3,700 miles (6,000 km) above the Earth. The outer belt reaches from 8,100 to 37,300 miles (13,000 to 60,000 km) and contains high-energy electrons. While the atmosphere blocks most high-energy particles from reaching us on the ground, the belts can be a danger for satellites, which need to have their electronics specially shielded if they spend much time in orbit there. NASA was fully aware of the belts and their potential as a radiation hazard. The narrower inner belt was avoided as the spacecraft departed Earth, while less dense regions of the outer belt were selected as exit points. Paired with the craft's great speed, the astronauts got out of harm's way in a hurry. The average radiation dose they received was equal to two CT scans of the head or half the dose of a single chest CT scan. They all came back alive and well. Kennedy's promise was kept.

Resources

- Jodrell Bank confirmation: jodrellbank.net/20-july-1969-lovell-telescope-tracked-eagle-lander-onto-surface-moon

- *Lunar Reconnaissance Orbiter* photos of Apollo landing sites: https://www.nasa.gov/mission_pages/apollo/revisited/index.html#.XHODh-JKi3c

- Many Apollo links to enrich your knowledge: https://www.history.nasa.gov/apollo.html

THE HUBBLE SPACE TELESCOPE CAN SEE THE APOLLO MOON LANDERS

+ ✦ +

No telescope on Earth can see the leftover descent stages of the Apollo Lunar Modules or anything else Apollo related. Even the miraculous Hubble Space Telescope can't do it. The laws of optics define the Hubble's resolution limit. The smallest thing its 94.5-inch (2.4-m) mirror can see measures just 0.024 arcseconds across in ultraviolet light. To appreciate how tiny of an angle that is, the full moon has an apparent diameter of 1,800 arcseconds, or close to ½°. That's how big it looks to our eye, but its true diameter is 2,160 miles (3,476 km). At the moon's distance, 0.024 arcseconds equals 141 feet (43 m). In visible light, its resolution is a bit less—0.05 arcseconds, or closer to 300 feet (91.4 m).

The largest piece of equipment left on the moon after each mission was the Lunar Landing Module, which measures 17.9 feet (5.5 m) high by 14 feet (4.3 m) wide. You can see the problem here. All signs of activity and equipment left by the Apollo program are unfortunately well below the telescope's resolution limit.

Hubble is optimized for solar system and deep space observing. The reason it can see details in everything from the Martian polar caps to distant galaxies is because all those objects are so much larger than the Apollo landers, even considering distance. Hubble also can take days-long time exposure photos to reveal the farthest, faintest things we can see.

For now, only NASA's *Lunar Reconnaissance Orbiter* (LRO), which can dip as low as 12 miles (20 km) from the lunar surface, gets close enough to image the landing sites, including the shadows cast by the flags the astronauts erected. To see those shadows from Earth you'd need a 650-foot (200-m)-diameter telescope.

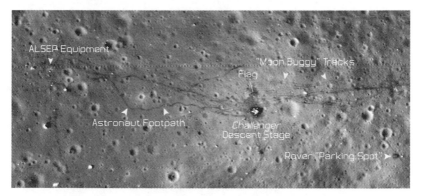

NASA's Lunar Reconnaissance Orbiter can orbit as close to the moon as 22 miles (35 km), giving it birds eye views of the Apollo landing sites. In this image of the Apollo 17 site, we can see the astronauts' footpaths, the lunar lander descent stage, equipment and even the shadow of the flag. (NASA)

Pluto (at left) covers only 15 pixels in the Hubble Telescope and shows almost no detail because of its small size and great distance. The Whirlpool Galaxy (right), which spans 60,000 light years, is 50 billion times farther away. But because it's also 250 trillion times larger, size trumps distance, allowing Hubble to get a clear and detailed photo even though the smallest thing it can see is 5 light years across! (NASA, ESA, S. Beckwith [STScI] and HHT [STScI/AURA])

Even the landers would require an instrument about 82 feet (25 m) across. The largest current telescope is the Gran Telescopio Canarias in the Canary Islands with a mirror 34 feet (10.4 m) wide, much too small for the job.

Even though the LRO's two narrow-angle cameras, those used for the lander images, have mirrors just 7.7 inches (19.5 cm) in diameter, the spacecraft makes up for its tiny telescopes by orbiting at a very low altitude. A big mirror is a great advantage, but closer is better. Much better.

ALIENS BUILT A SPACEPORT ON THE FAR SIDE OF THE MOON

+ ✦ +

Who hasn't stood in line at the grocery store and glanced at the cover of *National Enquirer Magazine* or the *Weekly World News* (back in the day) pasted with incredible headlines like "Dolphin Grows Human Arms," "Redneck Aliens Take Over Trailer Park" and "Adam and Eve Were Astronauts"? No one really believes these things, right? Most of us see these as a form of entertainment and a way to pass the time until checkout.

The same sort of titillating headlines and stories have migrated to social media sites in recent years. Pick a topic and you're bound to find "shocking new evidence" or apocalyptic predictions, including claims of alien rocket bases on the lunar far side. Could it be true?

We love this stuff because mystery matters. We live in a world where life can become mundane and predictable. Some people argue that science has taken the mystery out of the world because there's an explanation for everything (we're nowhere near that, BTW). Let me offer a different point of view.

While science does explain much about the natural world, it's constantly uncovering new, unanswered questions that fuel our curiosity to find out how, what and why. Through the efforts of scientists, we've discovered processes in nature no science fiction writer or poet would have dreamed of. You don't need to be a scientist to understand nature's ways, just a good observer. Nature shares its secrets when you pay attention. Science, or the scientific viewpoint, is all about finding the surprising connections between things that seem unrelated, like the flow of hot iron in the Earth's outer core and a compass needle, or the arrival of a meteorite from an ancient asteroid collision.

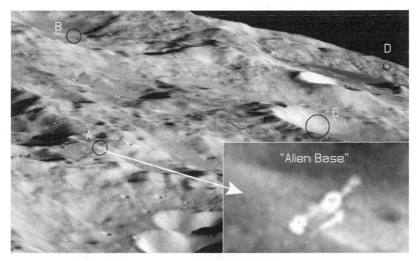

The so-called alien moon base (labeled A) is one of a number of dust flecks and film artifacts (B, C, D) in this photo taken by the Apollo 11 *astronauts. (NASA)*

Weird stuff riddles the online world to the point where it feels like a continuous game of whack-a-mole. A bizarre story appears, well-intentioned people attempt to debunk it and up pops another. And another.

The alien moon base is a great example. Photos of the lunar far side purport to show two interconnected launching pads for spacecraft. Because this is a pretty fantastic claim, we're going to need lots of evidence to prove it exists. That might include looking back over earlier photos taken of the area to see whether the base shows up.

Based on the known sizes of craters in the video image, the moon base is a multi-mile-wide structure. NASA's *Lunar Reconnaissance Orbiter* (LRO) has been photographing the lunar near and far sides down to about a 3-foot (1-m) resolution since 2009. Something that big wouldn't have been missed. Are there any current photos or pictures taken of its construction? Nope.

The only evidence offered in the video is that it *looks like* a moon base. And because it's on the lunar far side (which the video creator mistakenly refers to as the "dark side"; more on that myth later) and invisible from Earth, it must be a "secret" base. The entire premise is based on impressions and assumptions.

An alien base so close to home makes a wonderful story that instantly grabs your attention, but it's based on nothing but appearances, no hard data. What looks like a moon base to one person does not constitute proof that it is a base. In my city, we have a library built to look a little like a Great Lakes ship. If you saw it in a blurry aerial photo you might jump to the conclusion that it's a real ship, but no, it's a library.

The author did his homework in one regard. He provided the Apollo catalog number for the image, which anyone can look up. I did and gave it a close look. Not only did I find his lunar base but also half a dozen more! If that's what you want to call them. I call them dust and film defects, common in lots of space images, especially in the days of film, when this photo was taken. Each "moon base" is a sharp, white, opaque piece of dust resting atop the actual image, not a part of the lunar landscape. Such "artifacts," whether from film or digital processing, are among the most commonly misunderstood features in space images as well as online star atlases. Some are caused by light reflecting off the camera's lenses called *internal reflections*; they can assume a variety of shapes and sizes, making it easy for us to imagine them as alien spaceships or attempts by the government to cover up parts of the sky they don't want us to see.

Similar assertions of the moon being hollow and craters covered in glass domes are just that—fantastic claims based on either lack of information or a misinterpretation (deliberate or not) of a common occurrence. When the information is readily available and the person purveying the idea doesn't do their homework, they sow confusion among those who aren't familiar with the subject. That's a no-no and why the blob in the photo isn't a moon base. As for those glass-domed craters, they only look that way because of how they're illuminated by the sun. From certain sun angles, craters in a photograph can suddenly and stubbornly turn from concave, with sunken floors, to convex, with their floors bulging outward. It's only an optical illusion and well known to anyone who routinely looks at moon maps.

So many of these so-called anomalies are misinterpretations based on a lack of familiarity with the moon and lighting. Naturally, these rare or unusual sightings are inevitably "covered up" by authorities because the government wants them kept secret. Silence on a subject is nearly always interpreted as conspiracy, but the real reasons are more mundane—not to give credence to bogus claims by amplifying them to a wider audience.

I'm often surprised how quickly normally rational people will jump to the wildest conclusions to explain an "anomaly" before considering a simpler, more likely explanation. Maybe it's our inclination to consider the most human-threatening explanations first in case we might need to act to ensure our personal safely or that of loved ones. Such a catastrophic view adds unnecessary stress to daily life and clouds our intuitive good sense.

Whether through ignorance, deliberately ignoring the facts to get more hits or in the name of pushing a pet "theory," much disinformation has proliferated across the Internet. The result is predictable: those not familiar with the topic end up confused about what's true and what's not.

Resources

- Video of alien base: https://www.youtube.com/watch?v=s5uHk-mEJuMU&t=580s

THE MOON HAS A "DARK SIDE"

+ + +

It didn't start with Pink Floyd's *The Dark Side of the Moon*, but the popularity of that album added fuel to the fire. Since we only ever see the "bright side," we imagine that the moon's other half must be forever dark. Most people casually call it the "dark side," sometimes with a sly wink in a subtle reference to the *Star Wars* movies. So let's set things straight and get some needed light on the subject.

The side we see is called the *near* side. The side we don't see is the *far* side. Neither term includes a reference to sunlight or darkness, and for good reason— both experience day and night in equal amounts. Here's how.

The moon rotates on its axis at the same rate it revolves around the Earth, so it keeps the same face toward us at all times. If it rotated a little faster or slower, we would eventually be able to see the entire globe, but it rotates in lockstep with its revolution, turning at just the right speed to compensate for its revolution around the Earth. The far side is always turned away from us, facing away from Earth into outer space.

To get a feel for how the moon rotates, imagine walking in a circle around a pole. Your head is the moon and the pole represents the Earth. If you steadily face the pole as you circle it, your head will have rotated in a complete circle after one go-around. You're not aware of turning your head as you "revolve" around the pole, but you're doing exactly that. If you didn't turn your head to keep facing the pole, you'd be looking in different directions as you circled it.

From the pole's (Earth's) perspective, it sees only your face, never the back of your head, as you revolve around it. In exactly the same way we see just one face of the moon. This tight coupling of rotation and revolution is called *synchronous rotation* and is caused by the Earth's gravitational pull on the moon. We say that the moon is *tidally locked* to Earth because the Earth's gravity is powerful enough to raise tides within the moon's rocky crust, literally stretching and squeezing the moon as it

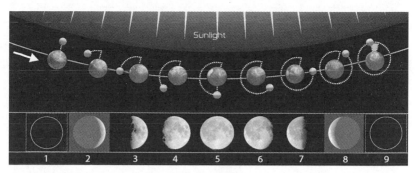

As the moon orbits about the Earth, the far side is fully exposed to sunlight and in full phase when we see a new moon (1); half-lit by the sun when the moon is half for us (3) and in complete darkness when we see a full moon (5). (Orion8 / CC BY-SA)

Both the near side (left) and far side of the moon get equal amounts of sunshine during each lunar orbit. (NASA)

spins. Friction within the moon caused by all this gravitational "kneading" gradually slowed its spin rate until it equaled its period of revolution. Think of it as the path of least resistance. Billions of years ago, the moon rotated much faster than it does today, so it was possible to see all sides of the orb during the lunar month.

The Earth-moon system is not the only example of tidal locking. Jupiter's four largest moons always keep one face toward Jupiter. Pluto and its largest moon Charon have similar masses and orbit so tightly that they're *mutually* tidally locked Charon faces one side of Pluto, and that side of Pluto faces Charon. On the far side of Pluto, Charon is never visible at all, while on the far side of Charon, Pluto is likewise absent.

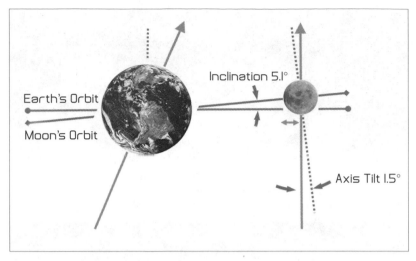

The moon's tilted orbit with respect to Earth's orbit combined with the 1.5° tilt of its axis lets us peek over its top and under its bottom, increasing the amount of surface we can see. This "nodding" is called libration in latitude. (Public domain / Wikipedia)

Just because we can't see the far side doesn't mean it never experiences daylight. As we learned, the moon rotates on its axis, so every part of it experiences both day and night. When the moon is new and near the sun in the daytime sky, we can't see it because the near side is turned away from the sun and in total darkness. We can only see the new moon during a solar eclipse when it slowly covers the sun. If you've ever seen an eclipse, you'll recall the moon looks black as night. There's a good reason for that—it's night across the entire near side! At the same time, the far side directly faces the sun. An astronaut orbiting the far side at that time would see a beautiful far side version of the full moon.

When we see a half moon in the evening sky, the orbiting astronaut sees the complementary half of the far side lit up by the sun. And when the moon is full on Earth, the far side is turned away from the sun and completely dark—that is, in new moon phase. As the near side goes through its phases, so does the far side but in the opposite order. See how easy that was?

Casual moon-watchers might think we see exactly one half, or 50 percent, of the moon. True enough. At any one time we see only half of the moon, but nature has been kind enough to provide an additional 9 percent thanks to a north–south wobbling and east–west rocking of the moon called *libration*.

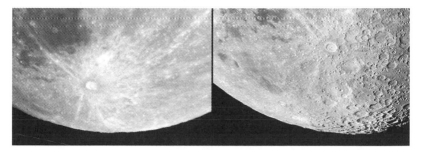

The nodding of the moon changes our perspective of the prominent lunar crater Tycho at the same time exposing extra territory along the moon's bottom (southern) edge. (Bob King [left], Frank Barrett)

To see the extra we'll need to observe the moon during the course of an entire lunar month. The wobble is caused by the 1.5° tilt of the moon's axis added to the 5.1° tilt of its orbit in relation to Earth's orbit. This allows us to look *under* the bottom of the moon during part of its orbit and *over* the top at other times. The side-to-side rocking occurs because the moon's speed varies as it revolves around the Earth in its elliptical orbit, exposing additional surface at its extreme east and west edges. Add up the snips and bits and we arrive at a grand total of *59 percent* of visible surface.

Thanks to space probes in lunar orbit, you don't have to become an astronaut to see the far side. NASA's *Lunar Reconnaissance Orbiter* (LRO) has mapped 98.2 percent of the moon's surface to a resolution of 328 feet (100 m) and in even greater detail in some regions. If you ever want to see the far side in all of its glory, go to the website listed in Resources and click on the LRO's large-scale map. Maybe you'll even want to dial up some Pink Floyd as you browse away.

Resources
- The far side of the moon: https://www.nasa.gov/mission_pages/LRO/news/lro-farside.html

FULL MOONS ARE A THING

+ ✦ +

Astronomers reckon the start of the moon's cycle, called a *lunation*, at the moment of the new moon. About two weeks past new, the moon lies opposite the sun in the sky, faces it squarely and shines big, round and full. As seen from space, the sun, Earth and moon line up in that order at every full moon. Full moons repeat every 29.5 days for an average of about one per month. Or so we've been taught. If you'll allow me to split hairs on exactly what a full moon means, we'll have a little fun and illustrate a point, too.

The precise alignment I just mentioned only really happens during total lunar eclipses, which occur about twice every three years. Partial eclipses and penumbral eclipses, where the moon passes through Earth's outer shadow, called the *penumbra*, occur more frequently. But only during a total lunar eclipse can the moon be said to be 100 percent illuminated and therefore a *full moon*. Only then do the sun, Earth and moon lie in a *straight* line. You can hardly miss the irony here: the moon is truly full only when passing through the Earth's shadow.

You can still easily see most totally eclipsed moons because of sunlight filtering around the circumference of the Earth that falls into the shadow and colors the moon in shades of red, orange and occasionally deep brown, depending on the clarity of Earth's atmosphere. Were it not for the atmosphere (which refracts sunlight into the shadow), a true full moon would be invisible to the eye, lost in inky shadow for a couple of hours before finally peeping back into the sunlight.

When not in total eclipse, the lineup is slightly askew and the moon not fully full. It might appear perfectly round to your eye, but if you train a telescope on it, you'll notice a small amount of shading along its northern or southern edge even at the exact moment of "full moon." That's because the sun isn't shining directly at the moon but at a slight angle.

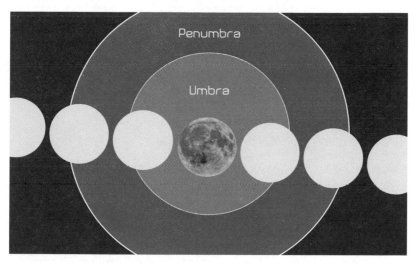

Only when the moon lines up behind the Earth and sun in a straight line can it truly be said to be full. Curiously, this only happens during a total lunar eclipse. (Tom Ruen)

During most full moons this never happens because the moon's orbit is tipped 5.1° to the plane of Earth's orbit. If the lunar orbit had no tilt—that is, if it circled the Earth in the same plane as Earth circles around the sun—it would pass into Earth's shadow at every full moon and we'd see an eclipse every month. Eclipses occur infrequently because most full moons pass above or below the plane and miss the shadow. The average number of partial and total eclipses in a year ranges from zero to three. The moon crosses Earth's orbital plane at other phases, too. When it happens at new moon, we see an eclipse of the sun.

Getting back to the moon's appearance in a telescope, when the "full moon" lies north of the plane of Earth's orbit, we see shading around its northern edge. When it's south of the plane, shadowing is visible along the southern edge.

The next time you hear the moon is full you can chime in with "close, but no cigar." Or you can just keep it to yourself and not annoy your friends. But now that you know the facts, you can't unknow them. For that I apologize.

YOU CAN BUY LAND ON THE MOON

+ ✦ +

Free enterprise knows no bounds. That's no truer than in the business of selling outer space property where at least five companies are hawking land on the moon. Many don't stop there, having moved on to Venus, Mars and the moons of other planets. For $25 to $30 (2019 price) you can get a deed to an acre of the Sea of Tranquility. Need more room? A hundred bucks at one outfit will get you five acres and a deed in a frame with your name on it. All new property owners also receive a map showing the location of their purchase and assorted documents.

Sites boast of having sold lunar real estate to Hollywood celebrities, former U.S. presidents and even NASA employees. One firm claims more than six million "proud owners." I'll be honest, I had no idea so many of my fellow citizens own plots on the moon. Or do they?

Ownership is a delicate issue, as anyone who's inadvertently trespassed on someone's land soon discovers. Just as there are property laws on Earth, so too are there on the moon. Rules of exploration, ownership and the peaceful use of the moon and other celestial bodies are covered in the Outer Space Treaty, which was crafted by the United States and the then-Soviet Union. Both countries signed the treaty in 1967, as have many more nations.

If you take the time to read it you'll find this stipulation in Article II:

> Outer space, including the moon and other celestial bodies, is not subject to national appropriation by claim of sovereignty, by means of use or occupation, or by any other means.

Further, Article I states that "there shall be free access to all areas of celestial bodies."

In other words, a country like China or the United States can't take over or occupy any piece of the moon. Neither can they claim the moon—even a part of it—for themselves. In this general way, the treaty addressed land ownership on the moon in anticipation of humanity's inevitable arrival there. After the Apollo missions and Soviet unmanned sample returns, when going to the moon became a reality, a separate, more detailed "annex" was crafted by the United Nations to close loopholes. Named the Agreement Governing the Activities of States on the Moon and Other Celestial Bodies, it's better known as the Moon Treaty of 1979.

Article 1 states that everything in the agreement related to the moon "shall also apply to other celestial bodies within the solar system other than Earth." But the relevant parts for would-be landowners are found in Article 11. Stay with me as we wade through a smidge of legalese:

> Neither the surface nor the subsurface of the moon, nor any part thereof or natural resources in place, shall become property of any State, international intergovernmental or *non-governmental organization, national organization or non-governmental entity or of any natural person.*

Here's what's clear. No one—not you, not me, not any business on Earth—can legally own property on the moon, or on Neptune for that matter. Does that mean you can forget that cozy cabin you were planning to build on your lunar lot?

Maybe yes, maybe no. As of January 2018, only eighteen countries were parties to the treaty, and none of them had yet to send crewed missions to the moon. The United States might have signed on were it not for a strong government lobbying effort in 1980 by the L5 Society—a group originally founded to promote space colonies. Representatives successfully argued that the treaty would make space colonization impossible, restrict U.S. commercial interests and stymie future efforts at terraforming other worlds to make them more fit for life. In part because of the group's efforts, the United States never signed on. Nor did Russia, China, the U.K. and other spacefaring nations. The L5 Society later merged with the National Space Institute to form the National Space Society, which continues to advocate for human exploration and colonization of space.

That leaves only the original treaty as law when it comes to lunar property, and it's written to apply to "states" or independent countries. What's to stop mom-and-pop entrepreneurs from starting a lunar real estate business? Business owners read the treaty as forbidding *state* ownership, but not necessarily *private* commercial sales. We could argue that this narrow interpretation sidesteps the spirit and intent of the agreement, which focuses on sharing and using the moon for peaceful purposes for the benefit of humanity.

Does this mean that technically, people *can* sell land on the moon and other celestial bodies? Before you get too excited, there's one more wrinkle, this from Article VI in the original Outer Space Treaty (the one the United States signed):

> The activities of non-governmental entities in outer space, including the Moon and other celestial bodies, shall require *authorization and continuing supervision* by the appropriate State Party to the Treaty.

Because the U.S. government has not authorized claims and sales of lunar properties, deeds and all the rest have no legal basis. Therefore, no one except the organizations selling the land will recognize your claim, and no one can ensure and enforce your property boundaries. If Russia puts a lander there and starts drilling for rocks, you're out of luck.

On the other hand, no one is enforcing the treaty either, so businesses have free rein to sell moon real estate so long as someone is gullible enough to pay for it. Owning acres of moon dirt is exactly like paying money to name a star. No astronomer—beginner, amateur or professional—will recognize the name. It's strictly between you, the company and the person who received the gift, even if the certificate is signed in gold ink and framed in silver.

Maybe the folks who spend money to name a star or "own" a piece of the moon know in their hearts that no one will recognize their claim or use Grandma's star name in a scientific paper. For them, the sentiment is more important. Let's face it—we like giving unique gifts. Naming a star or "owning" the moon makes the receiver feel special. Every time the person looks up at the moon or in the direction of their star (most sold stars are faint and require a telescope), they feel a connection to the gift giver and perhaps to the universe at large.

But for me it rings hollow. All make-believe. If you'll allow, I have some different gift suggestions. How about a nice big star wheel to help learn the constellations or a small telescope and a map of the moon? With these you could explore the heavens for years like as if you owned the place.

THE MOON IS BRIGHT BECAUSE IT REFLECTS A LOT OF LIGHT

If you've ever walked under a soaring full moon on a winter's night, the landscape looks almost as bright as day. Normally, we can't distinguish color at night because of a lack of light, but under a high full moon, I can easily see the red in a stop sign and the green in my winter coat. The moon shines by reflected sunlight, and while it's nearly 400,000 times fainter than the sun, it still gleams so brightly we imagine it must be an excellent reflector.

If you're wont to walk at night, you'll first notice your shadow by moonlight when the moon waxes to a thick crescent. By first quarter phase, when the moon is half, it's very obvious. At the full moon, the sun illuminates the entire face of the moon; shadows are strong and dark, especially if they fall on snowy ground.

For a true blinded-by-the-light experience, try looking at a full moon through a 6-inch (15-cm) or larger telescope. You'll be taken aback by its brilliance. After you've had your look, that eye will be as useless as if someone had taken your picture with a camera flash.

But here's the crazy thing: If you were to pave a 2,160-mile (3,475-km)-wide asphalt parking lot—the diameter of the moon—let it age a few years and then put it in the moon's place, it would look just as bright. That's right. The moon is no brighter than an old asphalt lot or a conifer forest. The next time you pull into a parking lot, have a look around the lot—it's a moonscape out there.

A big, "bright" full moon rises over an ice-covered Lake Superior near Duluth, Minn. in Feb. 2019. (Bob King)

Astronomers measure an object's reflectivity by comparing the amount of light it receives to the amount it reflects back out into space, defined as its *albedo* (all-BEE-doh). Albedos range from 0 for absolute blackness to 1 for the whitest of whites. You'll most often see it expressed as a decimal or fraction of 1. The moon's average albedo is 0.12, meaning it reflects on average 12 percent of the light it receives from the sun back into space.

Not surprisingly, Earth's average albedo is 0.30 (about three times more reflective than the moon), brighter than green grass and very similar to terra-cotta roofing tile. If the planet were covered in conifer forest, its albedo would be 0.14, almost the same as that of the moon. If covered instead in ice, it would reflect back 84 percent of the light from the sun. Earth's higher reflectivity derives from clouds, snow, ice caps, deserts and, to some extent, the water that covers much of the planet.

Venus is the brightest planet because it's 100 percent covered in clouds, which excel at reflecting sunlight. That single fact boosts the planet's albedo to an amazing 0.76. If you could drag Venus in close enough to Earth to appear the same size as a full moon, the real moon would appear dull, dark and gray in comparison to the sheer white radiance of the goddess planet. We'd finally see the moon for the dim body it truly is.

Side-by-side comparison of 67P/Churyumov-Gerasimenko, a typical comet, and Saturn's moon Enceladus, the most reflective body in the solar system. (ESA CC BY-SA 3.0 [left], NASA JPL Caltech SSI)

As bright as Venus is, there's room for brighter. Saturn's moon Enceladus showers itself in fresh, bright ice crystals released by geysers from huge cracks in its surface called "tiger stripes." The geysers tap into a salty subsurface ocean; as the water makes its way from the warm core to the bitter cold of outer space, it freezes into a mist of ice crystals that coat much of the already icy moon. Enceladus reflects almost 100 percent of the sunlight it receives, making it the brightest object in the solar system.

Our moon more resembles a typical asteroid or the hard part of a comet called the nucleus in terms of reflectivity. The largest asteroid, Ceres, has an albedo of 0.09, a bit darker than the moon, while Comet 67P/Churyumov-Gerasimenko, studied by the Rosetta mission from 2014 to 2016, is darker yet, reflecting just 6 percent of sunlight it receives, about the same as dark, wet dirt. Halley's Comet, with an albedo of 0.03, is darker than charcoal (0.04). All this might make you wonder how it's possible to see these dimly glowing bodies in the first place.

Although charcoal and asphalt *do* look very dark compared to grass, buildings and homes, they stand out boldly against the black backdrop of outer space because they still reflect some light while space reflects none. It's all about context. For

Even very dark things like this charcoal briquette can appear bright when seen against an even darker background like a darkened room. Likewise can comets and dark-crusted asteroids. (Bob King)

example, a paper plate held against a snowdrift blends in, but if you hold the plate against a black background, it jumps right out. The same is true of comets, asteroids and our own moon. One might have a higher albedo than another, but none is as dark as space, so all look bright in comparison.

Night vision also plays into our perception of the moon's apparent brilliance. When we first step outside on a dark night, we can barely see anything, maybe one or two stars and the neighbor's light. But if we allow our eyes time to adapt to the darkness, we soon see well enough to find our way by starlight, moonlight and even local light pollution without assistance from a headlamp. Dark-adapted eyes not only let us see faint things more clearly, but they also make bright objects like the moon appear brighter still. If you've ever driven a long distance in the dark at night and then turned off the road to get gas, you may have felt your eyes wince at the brilliance of the lights in the station's canopy, something that would never happen in daylight.

It seems the world is always playing these little tricks on us, trying to confound our understanding. Our senses are the gatekeepers of our experience, and they're delightfully imperfect. Adding a dash of science to the mix gives us a clearer understanding of how we perceive the world.

BIRTH AND CRIME RATES SPIKE DURING FULL MOONS

+ ✦ +

"There must be a full moon tonight!" Whether you work at McDonald's or the *New York Times*, you've probably heard someone attribute a crime wave or bizarre behavior to the full moon. The last time I heard that line, I checked and the moon was in waning gibbous phase, several days past full.

I think we like to blame the moon for mayhem because we have no better explanation at the moment. Blaming the inanimate and uncomplaining moon has been a cultural catchall for seemingly forever. We pass it down generation after generation by repeating the anecdote until people start believing it's true. If any of us were to check the moon's actual phase during a wave of crime or a spike in births, we'd quickly discover that on most occasions, the moon wasn't full at all but a crescent, half or gibbous. The problem is no one checks. The announcement is made, people agree it must be so, then the next time something crazy happens, someone else repeats the old saw, and the myth perpetuates itself.

Add a dash of *confirmation bias*, the tendency to recall all the instances in which the evidence supports our belief and ignore all the evidence that doesn't, and you've got yourself a full-blown urban myth. For instance, we might recall when a big crime was committed during a full moon but ignore those times when similar crimes were committed at other phases of the moon—assuming we're checking phases regularly, which again, most people don't do.

At the heart of the lunar blame game is the age-old folk belief that the moon influences human behavior—the full moon in particular. It's possible it might influence us indirectly through the placebo effect, a brain game where if you believe it can affect you it will. In the modern era, when pressed as to how the moon could

Does the moon really affect birth rates and make people go loony? Naw, probably not. (Bob King)

possibly affect a person's behavior, some will point to the power of the moon's gravity. If it can raise tides in the oceans, it must also exert a pull on each individual.

They're absolutely correct, but before we jump to conclusions, let's look at the strength of the moon's gravitational pull. We can easily measure it using the known mass of the moon and Newton's law of gravity. When you do the math, it turns out that the moon's gravitational tug on a person weighing 220 pounds (100 kg) is the same as the tug of a 110-ton (100 metric tons) mass on that person from a distance of 3.3 feet (1 m). Because the average car weighs around 5,000 pounds (2,268 kg), the moon's physical influence on a typical person is equal to the tugging force of 44 cars in a dealership parking lot or the equivalent of standing next to a small building. The moon's gravity stretches your body—changing your height—by an amount 10,000 times smaller than the diameter of an atom.

A big city high-rise has far more gravitational might than the moon! We're just too tiny for the moon to mess with us. With the oceans, it's different. They're massive, and the moon's effect on them is significant. It's also important not to forget the Earth's gravity in all of this. We live on a big, massive planet compared to the moon with a gravitational pull six orders of magnitude stronger.

As for crime, you don't have to take my word that full moons have no effect on crime rates. Nearly every modern study on crime versus phase of the moon shows no connection. Crimes are as likely to occur at a full moon as they are at the first quarter and crescent phases. Other factors may influence when and how often crimes are committed, but full moons do not show a statistical difference over other phases.

If people associate the full moon with wild behavior or crime, others, including some nurses, doctors and paramedics, are convinced more babies are born at a full moon than during other phases. Once again, people generally don't check the moon's phase before making this assumption. Confirmation bias is also at play. You won't be surprised to learn that modern studies show no cause and effect between the moon's phase and birth rates. Check the resources below for details.

That said, full moons aren't without measurable effects on humans. I've observed that rising full moons have the power to coax out photographers by the hundreds even in the coldest weather. People will stop what they're doing to watch, exclaim and raise a phone to capture images and video of that big, orange orb climbing up the eastern sky.

Resources

- "Bad Moon on the rise? Lunar cycles and incidents of crime": https://www.sciencedirect.com/science/article/pii/S0047235210000589

- "Relationship between lunar phases and serious crimes of battery: A population-based study": https://www.sciencedirect.com/science/article/abs/pii/S0010440X09000030

- "Trauma and the full moon: A waning theory": https://www.annemergmed.com/article/S0196-0644(89)80014-9/abstract

- "The influence of lunar cycles on frequency of birth, birth complications, neonatal outcome and the gender: A retrospective analysis": https://www.tandfonline.com/doi/abs/10.1080/00016340802233090

- "Lunar phase and birthrate: A 50-year critical review": https://journals.sagepub.com/doi/abs/10.2466/pr0.1988.63.3.923

MOON PHASES ARE CAUSED BY EARTH'S SHADOW

✦ ✦ ✦

Lunar phases are the one thing even the most casual skywatchers are familiar with. Sometimes people wonder whether it's the Earth's shadow falling on the moon that causes the changing phases. At first blush, this sounds plausible because the Earth, being a sphere, naturally casts a curved shadow. But think for a minute—the shadow always falls *behind* the planet in one direction. And the moon is always on the move. Ceaselessly orbiting Earth, it spends precious little time in its shadows.

When you look up at the stars at night, Earth's shadow fills the entire sky. But like your own shadow, the planet's shadow narrows the farther away it falls. Seen from the side, it looks like a tapered cone, broadest on the night side but focusing down to a small spot with increasing distance.

At the moon's average distance of 239,000 miles (385,000 km), the width of Earth's dark, inner shadow, called the *umbra*, is about 5,600 miles (9,000 km) or about 2.6 times the moon's diameter. That's equal to 1.5° or about the width of your index finger held at arm's length against the sky. That's a pretty small target!

For the moon to pass into the planet's shadow, it must stand almost directly behind the Earth in line with the sun. This is exactly what happens during a lunar eclipse and the reason lunar eclipses are the only times we can see Earth's curved shadow over the moon's face. The rest of the time what looks like a shadow over this or that half of the moon is nothing more than darkness. Lunar night.

The same way half the Earth is in darkness and half in sunlight applies to the moon, too. It's still night in that dark portion because the sun hasn't risen yet. The line separating day from night on the moon is called the *terminator*. If you keep an eye on lunar phases, you'll see the day-night line expand to the left (in the northern hemisphere) between the new and the full moon due to the

Earth's inner shadow at the moon's distance is circle only about 1.5° wide, a tricky target to hit, the reason lunar eclipses are infrequent. (Bob King)

ever-changing angle between the Earth, moon and sun. It's fun to watch the terminator change from concave (curving inward) at the evening crescent phase to a straight line at the half moon and then convex (curving outward) from the gibbous to the full moon. From new to full, the terminator exposes a fresh section of lunar landscape to sunlight night after night. After the full moon, the angle between the Earth and the moon decreases and the terminator runs in reverse, from convex to concave, marking the advancing line of sunset and the coming of lunar night.

The curve of the terminator is a natural consequence of how light falls on a spherical body. To see it yourself, get a ball, set it on a table and dim the lights. Then take a flashlight and shine it off to the right side of the ball to recreate a half moon. Hold it directly in front for a full moon and off to the right side for the last quarter. Crescents are trickier and require an assistant to go around almost to the back of the ball and shine the light back toward you to illuminate its edge.

My favorite lunar phase demonstrator is the Doppler weather radar ball that stands near a busy road in a neighboring town. When the sun is out and depending on my direction of travel, the ball can appear in any "phase"—from crescent to full. One time I deliberately drove around to the back and parked in the ball's shadow to create my own personal total solar eclipse. You'll find astronomical principles at play in the strangest places.

THE MOON PULLS THE OCEAN BENEATH IT, CAUSING THE TIDES

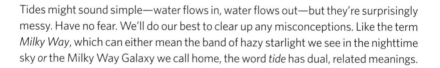

Tides might sound simple—water flows in, water flows out—but they're surprisingly messy. Have no fear. We'll do our best to clear up any misconceptions. Like the term *Milky Way*, which can either mean the band of hazy starlight we see in the nighttime sky *or* the Milky Way Galaxy we call home, the word *tide* has dual, related meanings.

When most of us hear the word *tide* we envision the twice-daily high and low tides in the ocean, familiar to anyone living along a coast. But there's also the *tidal force* which is the gravitational pull extended by one celestial body on another. This force not only raises water tides, but "landtides," too, flexing and slightly deforming the body's crust.

Although the sun has a much stronger gravitational pull on the Earth than the moon does, the moon's tidal force is twice that of the sun. That's because tides are created by the *difference* in gravity's pull on opposing sides of the Earth—that is, the side nearest the moon and the one opposite it.

The side of Earth facing a massive body like the moon feels a stronger gravitational tug than the opposite side 4,000 miles (6,437 km) away. Even though the moon is far less massive than the sun, it's so much closer to the Earth that the difference in the gravitational force it exerts between the front and back of the planet comes to 1.7 percent. That sun is so much farther away that the difference front to back is only 0.005 percent, or 340 times less. Clearly, the moon is the gravitational kingpin in the neighborhood.

That doesn't mean we should discount the sun. At the new and full moon, when the Earth, sun and moon all lie in a line, the solar pull reinforces that of the moon to create especially high tides called *spring tides*, named for the German word

The dramatic difference between low and high tide—some 20 feet (6.1 m)—is evident in this pair of photos taken at Chester Basin, Nova Scotia. (Peter J. Restivo)

springen, "to leap." When the moon is in the first and last quarter phases and the sun and moon are perpendicular to each other, the forces partially cancel each other out, resulting in lower than normal tides called *neap tides.*

If you've ever seen diagrams showing how the moon creates tides, you may have been left with the impression that the moon's gravity pulls the ocean below it into a big pile of water. That's not actually what happens. The moon is only ⅟₈₀th as massive as the Earth with far too little gravitational might to accomplish this feat. What it *can* do is pull water along the sides of the Earth through a type of gravitational attraction called the *tractive* force. Water pulled this way heaps into a big pile called a *tidal bulge* almost directly beneath the moon. As the Earth turns under the bulge, areas near the seas and oceans experience higher than normal water levels called a high tide.

You and I use traction when moving a super-heavy box from one side of a room to the other. Instead of lifting it up and risking a back injury, it's much easier to *slide* the box along the floor. The moon can't directly pull water upward to create the tides; the best its weak gravity can do is to "slide" the water over. Make sense?

That explains the bulge and tide on the moonward side, but what about that weird bulge on the opposite side of the Earth? How does that form? The strength of gravity weakens with distance. Double the separation between two bodies and it decreases by a factor of four. Because the center of the Earth is closer to the moon than the far side, the center is pulled *toward* the moon and away from the far side. This is exactly the same as having the far side (which feels a weaker tug from the moon because it's farther away) move *away* from the center of the Earth. Water responds by bulging outward on the opposite side of the Earth the same way it does

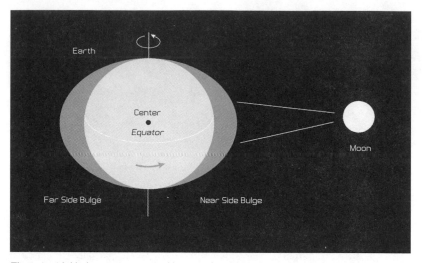

The twin tidal bulges are not caused by centrifugal force but by the moon's gravity. Even if the Earth stopped spinning the bulges would remain. The moon puts a squeeze on the Earth akin to squeezing a round balloon until the sides bulge out. (Bob King)

on the moon-facing side. So we get two bulges, one pulled toward the moon and the other away. You'll often read that the second bulge is caused by the centrifugal force of the spinning Earth, but this is incorrect. It's all the moon's doing.

The overall effect of the tidal force is to stretch an object. Because water is free to move, we get bulges, but the solid body of the Earth flexes, too, rising and settling about 1 foot (30 cm) a day.

When a location on Earth passes through a bulge, it experiences a high tide. Two bulges mean two high tides each day separated by two low tides between the bulges. The time from one moonrise to the next is about 25 hours, so high tides occur about once every 12½ hours. Water rises at a beach during high tide, and then drops to a low 6 hours later at low tide. Six hours after that the next high tide sweeps in, followed by the next low tide. This happens every day no matter the phase of the moon. As described earlier, the sun also plays an important though lesser role in tide making.

In the open ocean, the tidal force of the moon raises water about 3 feet (1 m), but that height varies a lot at any particular location depending on the angle and depth of the seabed, the shape of a beach and the strength of the prevailing winds. Because the moon's distance from the Earth varies around its orbit, its pulling power does, too. When closest, it raises the tides even higher. If that time coincides with a full moon, when the sun's share kicks in, then locations experience extra-high tides called *perigean* tides. If you want to experience the highest tides on Earth, head to the Bay of Fundy in Nova Scotia, where under the right conditions the water can rise up to 53 feet (16 m)!

Tidal forces between the Earth and the moon are responsible for both the slowing down of Earth's rotation and the fact that we only see one face of the moon. Earth's faster rotation carries the moon-facing bulge a little ahead of the Earth-moon line. The hump has mass and therefore gravity and pulls the moon forward in its orbit. At the same time, the moon's gravity pulls back on the hump in a sort of gravitational tug-of-war. The friction generated between the water and the solid Earth gradually slows the Earth's rotation.

Tidal friction is the reason the International Earth Rotation and Reference Systems Service has to add a leap second to the day every so often to take into account Earth's slowing rotation. In a hundred years the day will be 2 milliseconds longer. Based on studies of shoreline sediments called *rhythmites*, 620 million years ago a day on Earth lasted about 22 hours. In the distant future our descendants will experience longer days and hopefully more free time.

Meanwhile, the energy that Earth loses through tidal friction gets channeled into expanding the moon's orbit, causing the moon to creep away from our planet at an average rate of 1.5 inches (3.8 cm) a year.

Billions of years ago, the moon rotated much faster than it does today. Observers back then would have seen all sides of our satellite as it happily spun around. But over time, the friction generated by those twin rocky bulges within the lunar crust slowed the moon down until the bulges lined up pointed in just *one* direction—in line with our planet. The moon became *tidally locked* (basically the path of least resistance) with a rotation rate exactly equal to the time it takes to orbit the Earth. That's why we see only one side.

Gravity is always a mutual thing between one or more bodies, but especially so in the Earth–moon system, where the moon is relatively large compared to the Earth. Clearly, we've been in each other's gravitational hair for billions of years.

PLANETS, COMETS AND ASTEROIDS

THERE'S A PLANET ON A COLLISION COURSE WITH EARTH

+ ✦ +

If you're looking for a dramatic story for which there's not a shred of scientific evidence, let me introduce you to Nibiru, the online meme that refuses to die. Nibiru is a hypothetical planet on a collision course with Earth that first appeared in popular culture in the book *The 12th Planet* by Zecharia Sitchin published in 1976. The premise of the book is that an alien civilization called the Anunnaki came to Earth from Nibiru hundreds of thousands of years ago and genetically engineered the first human beings—the ancient Sumerian civilization of the Middle East.

Their planet is four times the size of the Earth, lies well beyond Pluto and orbits the sun once every 3,600 years, according to Sitchin's unique interpretation of ancient Sumerian myths and art. Further, Nibiru returns to the inner solar system every 3,600 years, when the Anunnaki set forth in their spaceships to visit the Earth.

Sitchin's books were and remain incredibly popular, with millions of copies sold worldwide and many passionate followers. Some believers claimed that Nibiru, also called Planet X by some, would return in May of 2003. It didn't. The fervor died down, but as is typical with these fantastic stories, it resurfaced in 2012, when it became linked to the supposed Mayan long-count calendar prediction of a world-ending cataclysm on September 23, 2017. But once again Nibiru was a no-show. Since then, people have conflated Nibiru with everything from a "second sun" to airplane contrails to Comet Elenin in 2011, a tiny object on an entirely different orbit that ultimately disintegrated as it approached the sun.

A planet, an asteroid or a comet on a collision course with Earth has become a recurring theme in popular culture. I get the appeal. Fear of impending doom is

Imaginary scene depicting a malevolent Nibiru attacking the Earth. (PerisLagios1999 CC BY-SA 4.0)

built into our DNA. We've survived as a species this long in part because we're good at imagining catastrophe. Better to fight or flee than be caught by surprise. We also love a good story.

Let's examine the facts. The word *nibiru* is real. That much is true. It's an ancient Sumerian term that meant "crossing" or "crossing point" when used outside of astronomy. But in connection with the sky, Nibiru had multiple meanings—the planet Jupiter, Marduk the chief god of Babylon, Mercury and generically as "a star." It's possible that the name Nibiru was the name given to a planet at a key celestial crossing point in the sky like an equinox or solstice. Sitchin instead interpreted "crossing" as Nibiru *crossing* the orbits of the other planets as it traveled from a great distance away to the inner solar system. Okay, but let's be honest. That's a stretch. Hard numbers, please.

Like all civilizations before the invention of the telescope, the Sumerians knew of only five planets in the night sky: Mercury, Venus, Mars, Jupiter and Saturn. Because they referred to Nibiru as one of the familiar planets, it's clear they weren't hinting at another planet beyond those already known. Records clearly show that Nibiru was also seen every year, not once in 3,600 years. What about the Anunnaki? They were important mythological deities who decided the fate of mankind.

Throughout history, gods have been endowed with superpowers not available to mortals. In that regard, the Sumerian gods don't stand out as extraordinary. Gods creating humans also isn't a new idea, but ancient astronaut gods that routinely visit Earth from another planet is a new take on an ancient idea that doesn't stand up to scrutiny.

First there's the disconnect with the Sumerian understanding of the word, but on a more practical level, a big planet that routinely crossed the inner solar system would soon destabilize the orbits of the other planets after only a few revolutions around the sun. Similarly, gravitational shoves from those same planets would have altered and changed the object's orbit.

When Nibiru was thought to pass near Earth in 2003 and 2012, it would have been easily visible with the naked eye as a slowly moving "star" by the time it reached the orbit of Mars. Nothing was seen. Even if Nibiru were somehow invisible, astronomers still would have detected its effects on the orbital motions of the other planets. From that, its position could be determined and a telescope pointed to it.

Some have said that NASA is covering up the planet's existence, but even if true, it wouldn't matter. While the agency has astronomers on its payroll, there are hundreds of thousands of amateur and professional astronomers across the planet who look at and photograph the sky every night. Not to mention all the professional astronomers at universities with no connection to the government or space agencies independent of NASA, such as ESA (European Space Agency), JAXA (Japanese Aerospace Agency), Russia's Roscosmos or the China National Space Administration (CNSA). No planet would escape this net of skywatchers.

When it comes to spectacular claims and supposed NASA cover-ups, what should NASA or professional astronomy organizations do? When the claim has no scientific validity, the agency risks giving more exposure to a pseudoscientific idea. If it remains quiet, some people interpret that as further evidence of a cover-up. It's a tough position to be in. Occasionally, NASA will go out of its way to address concerns that get out of hand. In 2012, Don Yeomans, a senior research scientist at NASA's Jet Propulsion Lab, posted a YouTube video to allay fears of Nibiru in the context of the Mayan calendar prediction. His quiet, thoughtful approach settled some nerves, but the whole incident, based on fear or a fiction should never have happened in the first place.

Astronomers have long considered the possibility of planets beyond Neptune, even the hypothetical Planet Nine, but all of these candidates follow orbits that lie well beyond Neptune and never approach the Earth. The only objects we're aware of that drop into the inner solar system from Nibiru-like distances are comets from the Oort Cloud, located between 2,000 and 100,000 times the Earth's distance from the sun.

Comets are small, icy objects typically just a couple of miles (km) across. Far from the sun, they hover near absolute zero, with surface temperatures colder than -400°F (-240°C) and no atmosphere, an unlikely place for a race of superior beings to set up a civilization. Not to mention that their world would partially vaporize as it approached the inner solar system and began to react to the increasing heat from the sun.

If Nibiru is starting to sound made-up to you, you're listening to your inner skeptic. All of us have a *baloney detector*, an internal reality check we consciously or unconsciously use to measure whether we're being told the truth or not. Call it a sixth sense. Sometimes our detectors get a little rusty, especially when we're unfamiliar with basic scientific principles. This can make us easy targets for those who might exploit our ignorance, whether to gain notoriety, sell a book, incite fear or simply out of their own ignorance of the facts.

The Internet's a wonderful place for looking up facts and doing research, but it's also booby-trapped with misinformation and pet "theories" that have no scientific basis. Anyone sharing science-related information owes it to their audience to double-check the facts before presenting their ideas as science.

There's nothing wrong with having an opinion and making speculations, but readers and video viewers should know where the facts end and imagination begins. Newspapers have clearly marked editorial pages so the reader knows when they've entered the land of opinion. To those who have a pet "theory," surmise, hypothesize and postulate to your heart's desire, just allow that you may be completely incorrect. Be frank with readers that yours is an interpretation and lacks hard, verifiable evidence. When it comes to astronaut gods from undetectable planets, we need proof by the bucketful.

WE FOUND BIGFOOT ON MARS

+ + +

I have the uncanny ability to see painterly portraits of men and women within the patterns of bath towels and tile floors. What, you too?! As my eyes relax, a faint outline appears that grows more detailed as imagination fills in the blanks. In seconds, a masterpiece materializes before me.

If you too have imagined faces in all the wrong places, you're under the spell of pareidolia (pair-eye-DOLE-yuh), the human tendency to see a familiar image or pattern where there is none. It happens all the time, with faces and animals being most common. Recognizing the human countenance is hardwired into our brains from birth. Think how important it was to find a friendly face when it came to survival. Still is. Pattern recognition also helped us find food, seek shelter and locate a mate. It's what makes Rorschach inkblots a useful tool: What we project onto an abstract pattern may tell us something about ourselves.

No one would ever confuse those towel-y faces for something real, right? Well, not always. When it comes to photos taken of unfamiliar places like Mars or the moon, it's easy to let our imaginations stray. That's why some people (see Resources) saw Bigfoot and other humanoid and animal forms prowling about the rocks in a photo taken by NASA's *Spirit Rover* on January 25, 2008. You and I might agree on the striking similarity of rocks to animals, but a few people went a step further and claimed what they saw were real, living things.

Whoa, that's a big leap. Let's pause to consider several crucial facts. First, Bigfoot's size in the photo. Using AlgorimancerPG software (see Resources) for determining distances and sizes of features in Mars' rover images, we can determine that that the odd rock is about 15 feet (4.6 m) from the rover, making it at most about 2.4 inches (6.1 cm) tall. OK, maybe it's an itty-bitty baby Bigfoot, but that's REALLY small. Wouldn't you agree it's far more likely to be a weirdly

A remarkably lifelike "Bigfoot" strolls across the Martian landscape in a photo taken by NASA's Spirit Rover in 2008. Too bad he's only about 2.4 inches (6.1 cm) tall and made of rock! (NASA / JPL-Caltech)

eroded rock? Dusty winds have been whipping the Martian sands around for the past several billion years.

Here's another reason it's probably not alive. When the rover takes a color picture, it doesn't do it all at once but shoots three pictures through blue, green and infrared filters, and then combines them into a single color image. The process takes more than a minute, during which time Bigfoot probably would have moved. I mean, he *is* in classic stride-mode.

Still not convinced? Consider the weather at the time. *Spirit* explored Gusev Crater and the surrounding region, where the daytime temperature averaged between -50°F (-45°C) and 23°F (-5°C). I'll admit that it's not a deal killer, but the atmosphere is 100 times thinner than Earth's and comprised of 96 percent carbon dioxide. Any mammal would choke for air, feel its blood boil, quickly pass out and then freeze hard.

It's safe to conclude, based on easily available information combined with the knowledge that we can't help but read patterns into randomness, that "Bigfoot" does not inhabit Mars. Probably not the Earth either, but that's a discussion for another day.

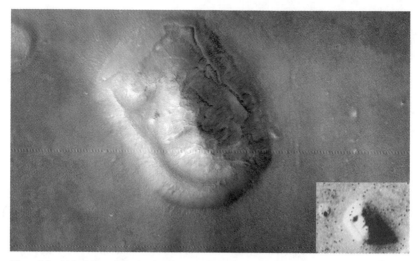

The original Mars face (inset), photographed in 1976 by NASA's Viking 1 *spacecraft resembles a human face. The higher resolution image taken by NASA's* Mars Reconnaissance Orbiter *better reveals the face's true nature as one of many eroded hills on the Red Planet. (NASA / JPL-Caltech)*

Other factors like poor lighting, lack of detail and fuzziness in photos can lead our suggestible selves astray. Go to YouTube some time and look up "Mars alien." You'll see dozens of videos based on Mars rover photos that purport to show aliens lurking in the shadows, a Sasquatch skull and even a humanoid thumb. The Red Planet has a peculiar hold on our imagination.

Viking 1 spacecraft images taken from Mars' orbit in 1976 at low resolution (the best at the time) seemed to show a human face carved in rock staring back at Earth. Located in the Cydonia region of hills, mesas and knobs, the sight inspired some organizations and individuals to see it as evidence of intelligent life. Before long, the dusty mesa was dubbed the "Face of Mars" and imagined as the work of Martians, who also built "pyramids" and other structures found in the vicinity.

Members of NASA took a cooler approach, explaining that poor resolution and lighting were to blame—along with our propensity to see faces in random patterns—and that future missions to the Red Planet with better cameras would whisk away the illusion.

That's exactly what happened. The high-resolution camera on the current *Mars Reconnaissance Orbiter*, which can see details as small as 11.8 inches (30 cm) across from an altitude of 186 miles (300 km), clearly shows a 2-mile (3-km)-long eroded hill. What tricks of lighting made easy to see as a face disappeared into a maze of eroded details once a better camera was used.

Yeah, I wish it were the work of Martians, and beliefs die hard—some people still accuse the space agency of conspiracy on this one—but we know from our own earthly experience that our senses can deceive us. We want to trust them, but not so much that we sweep logic and fact aside.

Remember this when confronted with an incredible story or impossible photo, especially when it appears on social media—*extraordinary claims require extraordinary evidence*. The more bizarre the notion, like Bigfoot roaming around Mars, the more proof you'll need to convince us it's true. Merely looking like something isn't good enough. Carl Sagan, the late American astronomer and astronomy popularizer, made that phrase popular. I like it because it keeps us on our toes.

Resources

- American lawyer discovers life on Mars: http://www.prweb.com/releases/2009/01/prweb1803034.htm

- AlgorimancerPG application download: http://www.clarkandersen.com/RangeFinder.htm

COMETS CAUSE EARTHQUAKES

✦

Some of you may recall seeing Comet Hale-Bopp ablaze on spring evenings in 1997 or Comet McNaught with its magnificent, peacock-like tail splayed across the 2007 summer evening sky in southern latitudes. If not, they're worth a Google. A tail is a comet's most iconic feature, giving it the appearance of a giant fireball suspended in space. Comets looked scary, one of the reasons they rattled our ancestors. Even worse, they appeared unexpectedly, literally out of the blue, unlike the predictable stars and planets. Unexpected and generally unwelcome, at least in European history, comets were blamed for pestilence, fires, deaths of kings and bad fortune in battle, among other things. Echoes of how we still blame the moon for high crime and birth rates.

The bright tail we normally see with the naked eye or binoculars is composed of fine dust about the same size as the particles in a cloud of cigarette smoke. The dust scatters sunlight and glows like dust caught in a sunbeam shining through a window. A typical tail ranges from 600,000 to 6 million miles (1 to 10 million km) long. Comet Hyakutake, which put on a fabulous show in the spring of 1996, is the current record holder (as of early 2019), with a tail that reached over 354 million miles (570 million km), or more than four times the distance between the Earth and the sun.

Get this: If you could get out there with a push broom and sweep all that dust into a pile, it would fit in an ordinary suitcase. That's how little material it takes to make a comet's tail. Dust impregnates water ice as well as ices of carbon monoxide, carbon dioxide, methane and ammonia that comprise the solid part of a comet called the *nucleus*. A comet nucleus is neither very large, nor dense and rocky like an asteroid, but more like a dirty, rather fragile snowball honeycombed with hollows. I like to compare it to that black, gunky snow that gets packed in your car's wheel wells during winter.

Comet McNaught, also known as the Great Comet of 2007, was a spectacular sight from the southern hemisphere in the winter of 2007. (ESO:Sebastian Deiries)

Dust is released when a comet comes near enough to the sun for the ice to vaporize. Sunlight gently pushes the tiny particles back behind the comet's head into a long tail. Comet nuclei diameters are notoriously hard to measure because they're well hidden by the dust and gases released by vaporizing ice. Our best data and photography come from a dozen spacecraft flybys and one orbital mission to seven different comets. Comet nuclei range in size from 37 miles (60 km) ± 7.5 miles (20 km) for Comet Hale-Bopp, down to about 1,050 feet (320 m) for P/2007 R5. Most are only a few miles (km) across.

The nucleus of Halley's Comet has a diameter of 6.8 miles (11 km). Standing on its icy, dust-laden surface, a 200-pound (91-kg) person would weigh about 0.16 ounce (4.5 g), equal to 1 teaspoon of sugar. Someone figured out that the total mass of Halley's Comet equals one Mt. Everest. Think what a minute fraction Mt. Everest is compared to the mass of the Earth. Or even the moon. Even big boy Hale-Bopp is just 1/500,000,000th the mass of Earth.

Moreover, every time a comet zips through the inner solar system, heat from the sun causes it to lose some of that mass through vaporization. During its most recent appearance in 1986, Comet Halley shed 30 tons (27 metric tons) of gas and 24 tons (21.9 metric tons) of ice *per second* at its peak! Although they are massive enough for many orbital cycles, over time, comets can simply fizzle away.

Mass and distance are what determine the strength of an object's gravitational pull. If an object is extremely massive and passes close to a planet, the planet will experience a strong attraction. But if it's close and tiny, attraction will be next to none. Comets are so miniscule compared to the Earth that their gravitational pull is insignificant. In fact, it's the other way around. A close-approaching asteroid or comet will fly in one way and leave on a slightly different track, its orbit altered by *Earth's* gravity. To date, the closest comets have passed within 1.1 and 3 million miles (1.8 to 4.8 million km) of Earth.

Unless a comet is on a direct collision course with our planet, we have nothing to fear from them. To the best of our knowledge, no known asteroid or comet is expected to strike the Earth for at least the next 100 years. When comets are many millions of miles from Earth, they're even less of a concern. While it's theoretically possible to measure the gravitational force a comet exerts on Earth from 50 million miles (80.5 million km) away, the pull is even less than what Mt. Everest exerts on you as you turn the pages of this book.

That's why if you ever hear someone say that a comet is responsible for causing an earthquake or a catastrophic weather event you can be sure it didn't and couldn't. This is a fact even when a comet lines up with another body and the Earth. Whether it's in alignment or not, the comet's gravity is the same. Because the comet's gravity is insignificant to begin with, the alignment will not enhance its pull. There's no magical force that comes into play during celestial body alignments. And although it's true that planets are far more massive than the biggest comet and have stronger pulls, they're also so far away that their gravitational effects on Earth are practically negligible.

Let's take the biggest planet, Jupiter. It exerts the strongest pull on Earth of any of the planets, yet that force is still only 0.0000068 times that of the moon! Venus is next, with 94 percent of Jupiter's pull, followed by Mars, with 41 percent. All the others are 7.4 percent or under. Even if it were possible (and it's not) to exactly line up all eight planets in a perfect line on one side of Earth for maximum gravitational pulling power, the force on Earth would still be microscopic. In contrast, the sun's gravitational force on Earth is 29 orders of magnitude higher, making all the other planets' gravities virtually irrelevant.

Earthquakes and bad weather have other well-known, Earth-centric causes. Earthquakes originate in the movement of rock in underground faults, while weather is ultimately caused by solar heating along with variations in air pressure, humidity and wind. If a comet makes big headlines at the same time an earthquake occurs, we're understandably tempted to connect the two. Simultaneous events or coincidences seem to imply a connection as if they share a common cause or one leads to the other. But in the grand scheme of things, coincidences occur all the time. Think how many things are happening simultaneously right at this moment. In fact, there are so many coincidences we can't spare our attention to notice. Instead, we tend to remember those that either involve dramatic events or carry some deeply personal meaning.

My mother passed away several years ago. Shortly *before* I received the news, I had turned down an offer to camp overnight with friends. Instead I drove home, and that's when I got the phone call that she had died. I was grateful to be heading home to my wife instead of telling stories around the campfire. Did I have a premonition that led to my choice? No. I left camp for other reasons. The two events simply happened about the same time. It would be tempting to believe that personal coincidences are guided by some invisible hand, but I don't believe the world works that way—period. Many would disagree, and I totally understand. But you see, I'm stuck with myself.

Once in a while there really is a connection between two things that occur closely in time, like lightning and thunder. We know what that means—seek shelter! Indeed, the human penchant for making associations may have contributed to our survival, but sometimes we overdo a good thing, another popular human trait.

False connection-making happened with a vengeance in 2011, when astronomers predicted that the recently discovered Comet C/2010 X1 (Elenin) would likely become bright enough to see with the naked eye. On its way toward the inner solar system, a serious earthquake occurred in Japan. People who should have known better attributed the tragic event to an alignment between the comet, Earth and another planet. Once we knew the shape and length of the comet's orbit, some people traced Elenin's path back in time and announced that earlier comet-planet alignments jived with additional earthquakes recorded in China, Peru, Indonesia and many other locales. Elenin was on a rampage! Before long, people were calling it the "doomsday comet."

But a *correlation between events doesn't necessarily imply a connection*. Here again, we made the mistake of assuming two things are related or that one causes the other because they happen at the same time. The same person who drew up the list of comet alignments compared them to the times of earthquakes and found many matches. Did the comet cause these horrific events?

Between August 19, 2011 (left) and September 2 Comet C/2010 X1 (Elenin) literally disintegrated before our eyes. (Michael Mattiazzo)

Let's say instead of earthquakes, we used the dates I did laundry. Assuming I do the wash about once a week, I could easily find earthquakes to match many of those dates because they occur with great frequency. Magnitude 2 and smaller quakes happen several hundred times a day. Major earthquakes (magnitude 7) occur more than once a month. In the end, my comparative study would find that nearly every time I did my laundry an earthquake occurred. Did cleaning my clothes cause the earthquakes? Here's another. Let's say that over the past three years sales of black socks had increased at nearly the same rate as the number of wildfires reported across the United States. Okay, you get the picture.

Sometimes people will cherry-pick data to make a connection, selecting only the dates that correlate with their observations. Clearly, this is also a no-no. It may work to reinforce your personal belief about a matter, but it's not believable.

So no, comets don't cause earthquakes. Neither do planets or alignments between stars and planets and all the rest. We can appreciate the appearance of a connection but know we're smarter than that.

A final note: In a wonderful irony, it was Comet Elenin that faced doom in the end. Shortly before making its closest approach to the sun, the comet broke into pieces and fizzled away.

COMETS STREAK ACROSS THE SKY

✦ ─✦─ ✦

Bright comets are rare, but the image they create lingers in the public imagination. The bright head and long tail make it easy to associate a comet with a meteor or "shooting star," which has a similar appearance. Meteors are actually closely related to comets. Most of the meteors we see, from the nightly random fare to eagerly anticipated annual meteor showers like the Perseids, are composed of a mix of sand-size grains and small, rocky debris shed by comets as they're heated by sunlight.

Rock and dust locked up in the ice are released into the comet's temporary atmosphere called the *coma*. The physical pressure of sunlight—it doesn't take much force to move tiny bits of dust in space—pushes them away to form a tail. Material shed by the tail and coma spreads along the comet's orbit and some of it burns up as meteors in Earth's atmosphere. Most are tiny grains that never reach the ground as meteorites.

There are other big differences between comets and meteors. For one, comets are much farther away, with distances comparable to the planets. Encke's Comet, which orbits the sun every 3.3 years, is a frequent visitor to the inner solar system. When closest, it can pass within 16 million miles (25.7 million km) of the Earth. Meteors without exception flare into view between 50 and 75 miles (80 and 120 km) high. Their lives are brief, a few seconds typically, before they vaporize in the atmosphere. Meteors equal to or brighter than the planet Venus are called *fireballs*. Some rival the sun in brilliance. A large meteor that explodes in the atmosphere, potentially showering the ground with meteorites, is called a *bolide*.

Meteoroids, the solid particles lost by comets and released when asteroids collide, strike the atmosphere at speeds ranging from 25,000 to 160,000 mph (40,200 to 257,500 kph) and become meteors. A meteor's life is so brief you hardly have time to tell the person next to you to look up before it's gone. Comets travel at nearly identical speeds as they orbit the sun, but because they're so much farther away,

Comet Hale-Bopp (left) and a bright fireball both appear to streak across the sky, but the comet is nearly static because of its great distance, while the close-by meteor zips by in seconds. (Bob King [left], Vidur Parkash)

they appear to move much more slowly across the sky. Similarly, a plane seen at a great distance crawls across the sky compared to one seen up close on takeoff.

A comet's speed varies according to its distance from the sun. The closer to the sun, the faster it moves. As seen from Earth, a comet's apparent motion or speed across the sky also depends upon its distance from us. In general, a close comet covers more sky each day than a more distant one.

I've watched distant comets barely budge over several nights even though they're powering along at tens of thousands of miles an hour. A typical bright comet might move 1° to 2° per day, equal to the width of your thumb held at arm's length against the sky.

Comet IRAS-Araki-Alcock was one of the closest, brightest comets to pass Earth in recent years. In early May 1983, it safely missed us by just 2.8 million miles (4.7 million km). Over three nights the comet covered 100° of sky at a peak speed of three full-moon-widths an hour, painfully slow for a meteor, but a drag race for a comet.

There's another key difference between comets and meteors—comets aren't gone in a flash. There are records of Halley's Comet from as far back as 240 B.C.E., the first confirmed sighting. Many others return again and again on orbits that can extend billions of miles into the outer solar system. Although small by planetary standards and whittled away by solar heat during each swing by the sun, they have enormous staying power.

COMETS WILL SHOCK YOU!

+ ✦ +

You hear of lot of theory making nowadays, whether on TV, in movies or in ordinary conversation. Someone usually begins with "I've got a theory," and then throws out their best hunch or guess. I'm not a purist, so I usually don't go around correcting people on their use of the word. But because the meaning of the word is relevant to this topic, let's look at what a theory is.

The National Academy of Sciences defines a theory as "a comprehensive explanation of some aspect of nature that is supported by a vast body of evidence" and "can be used to make predictions about natural events or phenomena that have not yet been observed." The Academy is a nonprofit and nongovernmental organization established by Congress and signed by President Lincoln in 1863 that advises the nation on issues of science and technology.

A theory is neither a guess nor a surmise but a constellation of proven *facts* hard-won through years of data gathering, experimentation and computer modeling. A theory explains known phenomena but also makes predictions about related phenomena we have yet to measure or observe and synthesizes it all into a new way to understand nature.

If a theory's predictions are off or new evidence is found that runs counter to the theory, scientists have a couple of choices. They can either modify the theory to incorporate the new data or if the conflict is too great, the theory may get a major modification or be superseded by a new theory. Science is self-correcting that way. Older theories like the Ptolemaic system that placed Earth at the center of the solar system were rendered obsolete by Copernicus and his sun-centered version. Kepler refined that model when he discovered that planets move around the sun in elliptical, not circular, orbits.

The backbone of many theories, especially those in physics and astronomy, is math. If you can mathematically prove a theory, it makes it far easier for skeptical scientists to buy into your radical new idea. New facts can alter and expand a theory but must be rigorously proven with observational data and backed by math. The more radically a theory departs from the norm, the more experimental proof the theorist needs to provide to convince other scientists. Scientists are an extremely skeptical lot and will happily tear holes in your theory at every turn, subjecting it to a barrage of never-ending tests. But if your results and predictions are confirmed time after time, the theory becomes that much more robust, like a superhero who gains power by absorbing the strengths of other superheroes.

Good examples of theories that have withstood repeated challenges and still stand tall to this day include the heliocentric theory (sun-centered solar system, 1543), the oxygen theory of combustion (1770s), plate tectonics (1912), quantum theory (early 1900s) and evolution by natural selection (1859). Scientists will debate the details but all these theories are accepted as fact. Each has led to countless new discoveries, predictions and brand-new syntheses of knowledge. They are among our greatest legacies.

That brings us to the electric universe theory, or simply EU. If you spend any time online, you'll come across videos by its proponents who believe that electricity is the primary force in the universe. Wallace Thornhill, one of the founders of the "concept," earned a degree in physics and electronics at the University of Melbourne in Australia but left mainstream science to pursue a vision of the universe based on electricity. In the EU view, ropes of electrical plasma coil across the cosmos, charging the stars with electricity like so many cosmic streetlights. Invisible electric currents thread the solar system and are responsible for everything from the birth of the planets to electrifying comets to blasting out the canyons of Mars with giant lightning bolts. Where all that electricity comes from is left unexplained.

There are four fundamental forces in the universe: gravity, electromagnetism, the strong force and the weak force. The last two have to do with subatomic particles and aren't relevant to our discussion. We're interested in gravity and electromagnetism, the force between electrically charged particles.

Although gravity is the weakest force at the scale of subatomic particles, it's the dominant one when it comes to the universe at large. It causes lunar tides, keeps the Earth and planets in orbit and is largely responsible for gathering up hunks of ice and rock called *planetesimals* to build the planets 4.5 billion years ago. Unlike gravity, electricity is a strong force at small scales (atoms and molecules) but weak across large distances, like those between planets and stars. True, satellites

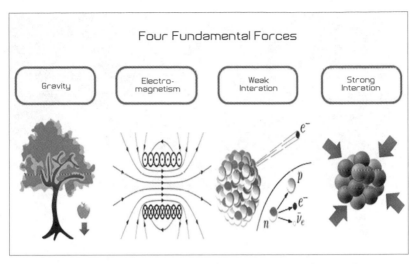

The four fundamental forces each has its own domain. Gravity extends across the universe while electromagnetism (and electricity) are more localized. The weak and strong forces operate at the atomic level. (Kvr.lohith CC BY-SA 4.0)

have detected powerful electrical currents in the upper atmosphere during major auroras, and Io's interaction with Jupiter's magnetic field connects the two bodies electrically, but these are local in the big picture.

EU puts electricity front and center as the governing force of the universe. One of its many claims is that comets are negatively charged and the solar wind, made of particles that stream from the sun, is positively charged, like opposite ends of a battery. As a comet approaches the sun and moves deeper into the positively charged electrical field, electricity from the positive-negative connection excites a comet to glow. At the same time, powerful *spark discharges* akin to static electricity drill into the surface, boring out tons of dust and gas while also turning the comet's nucleus the color of badly burnt toast. EU theory holds that comets are dry, electrically charged rocks. No water. No ice.

Not only have we studied comets in detail from Earth but also up close during spacecraft flybys and at least one orbital mission. For more than two years from August 2014 to September 2016, the European *Rosetta* orbited comet 67P/Churyumov-Gerasimenko. During its stay, it launched a small probe named *Philae* that bounced several times off the comet's surface (under very low gravity conditions) until it finally landed. EU followers predicted it would be zapped by electricity (remember, comets are supposed to be negatively charged) as

135

Jets of dust and gas shooting from Comet 67P/Churyumov-Gerasimenko originate from beneath the comet's crust and escape through fissures and holes in its surface. (ESA/ Rosetta/NAVCAM, CC BY-SA IGO 3.0)

the lander approached the surface. That never happened. The probe survived, performed experiments and transmitted data back to the mother ship before its batteries ran out of power.

The bright, fan-like array of what EU theory adherents call *electric discharges* are jets of dust and gas that erupt from below the surface when the comet is heated by the sun. As a comet's ice is warmed, it vaporizes into a gas, which finds its way to the outside through cracks or pits in the crust. Rosetta safely flew through one of 67P's many jets and emerged no worse for wear. The probe also collected dust from jets and photographed and analyzed it on-site. No sparking was detected, no electrical currents ever recorded.

The probe also found patches of ice on the surface and measured water leaving the comet's nucleus at a peak rate of several gallons per second. Yes, water! Comets have lots of it. We've seen it using ground-based telescopes and detected it directly at comet flybys and during the Rosetta mission.

As for 67P's charcoal-dark crust, it's composed of a layer of dust about 8 inches (20 cm) deep that partially insulates the ice below. *Rosetta* and *Philae* both detected carbon-rich organic compounds in the coma, some of which rain back down and accumulate on the surface. Multiple experiments have shown that when organic compounds are exposed to ultraviolet light from the sun and cosmic radiation, they darken. Because comets are rich in organics, it's no surprise that their exteriors are blacker than a penguin's back.

So, is EU a theory like quantum theory or plate tectonics? No. EU does not provide any mathematical analysis or experimental evidence to validate its claims. Neither has it been verified using the traditional scientific method. EU adherents examine current areas of research in science and propose fanciful, experimentally unverified ideas based on a single concept: electricity.

If traditional scientific models work, why would anyone go to the trouble of cooking up another theory to explain what's already been explained? Some people may like the one-size-fits-all aspect of EU. Every phenomenon, known and unknown, can be explained by electricity, a force we're familiar with that makes intuitive sense. Dark matter, relativity theory and the big bang are all just so much mumbo-gumbo to EU adherents.

Even though EU uses the language of science, it's really pseudoscience, a collection of beliefs that appear to be based on the scientific method but which are grounded in opinion, guesswork and unproven hypothesis. EU advocates, like other pseudoscientific enterprises, will often cherry-pick from a list of unsolved problems or "anomalies" and skewer traditional science for its inability to provide definitive answers.

Knowledge takes time. No one ever thought we'd know what the stars are made of, but then the spectroscope was invented, and now we know. We still don't know the origins of dark matter and dark energy, but (excuse the pun) I'd rather be in the dark for a while instead of getting an instant answer based on a one-size-fits-all "theory" with a poor track record for predictions.

There's a perception—often true—that scientists dismiss alternative "theories" proposed by non-scientists. If you're a scientist or science educator, sooner or later you'll be contacted by someone who has a revolutionary new idea that finally explains everything. But while the proposals often sound interesting and use scientific terminology, they read more like opinions with no scientific work to back them up. You can't deny the rare occasions when crazy ideas like relativity theory turn out to be the next scientific paradigm, but for every idea that hits a home run, a thousand are foul balls.

EU supporters undoubtedly love science as much as anyone, but it's not about believing in things that fit your perception of how the world should be, but something far more radical—the unexpected places new discoveries take us.

MARS APPEARS AS BIG AS THE FULL MOON WHEN CLOSEST TO EARTH

+ —+— +

In 2003, an Internet post appeared, claiming that Mars would look as big as the full moon when it made its closest approach to the Earth on August 27 that year. Mars and Earth pull alongside one another on the same side as the sun about once every two years at a time called *opposition*. Because the two planets are closest then, Mars shines especially brightly in the night sky. Mars' distance from the sun varies due to its elliptical orbit, so opposition distances vary too. The closest ones occur every fifteen to seventeen years.

There was a lot of excitement that August because at that particular opposition Mars would be only 34.65 million miles (55.76 million km) from Earth, the closest it had come since September 24, 57,615 B.C.E.—nearly 60,000 years ago! At a typical close opposition, Mars is about a million miles (1.6 million km) further from Earth and only slightly bigger and brighter. Astronomical rarities perk up our ears, and this one had the Internet buzzing.

At about the same time, someone shared an email describing details of the planet's close encounter with the Earth that unwittingly set a false rumor into motion. Here's the key snippet:

> At a modest 75-power magnification Mars will look as large as the full moon to the naked eye. Share this with your children and grandchildren. NO ONE ALIVE TODAY WILL EVER SEE THIS AGAIN.

The relative sizes of the moon and Mars seen from the Earth. You'd have to magnify Mars at least 75 times to expand it to the apparent size of the full moon. (Bob King [left]; NASA, ESA, and The Hubble Heritage Team [STScI:AURA])

Notice the line break after the word *magnification*. Innocent as it appears, it may have played a small but important role in misunderstanding the writer's intentions. Whoever wrote it described how Mars would look through a telescope *magnifying 75 times*. At the time, Mars measured 25 arcseconds across. That's a little less than one-thirtieth the apparent size of the full moon—large for Mars but still much too small for the naked eye or even binoculars to distinguish as a disk. But if you enlarge the planet 75 times in a telescope, its apparent size expands to the same as the full moon. *Apparent* in astronomy means how big something *appears*. If you somehow could look at Mars with one eye through the telescope and the full moon with the other eye, they would appear the same size. Not an easy thing to do, but possible, though I don't know of anyone who tried it.

Then something unfortunate happened. As the email made the rounds of the Internet, the reference to "at a modest 75-power magnification" was left out (perhaps in part because of the line break), leading neophyte skywatchers to expect a frighteningly large, moon-size Mars glaring down on Earth like a vision of the apocalypse. I remember getting frantic emails at the time asking me when to look to see this incredible apparition.

Because Mars is about twice the size of the moon, to appear as big it would have to be dragged from its orbit and moved to a spot about 480,000 miles (772,500 km) from the Earth. I'll leave that to your imagination.

Once Mars parted from Earth that fall and faded, so did the rumor. But like a dog that won't stop barking, the message popped up again across the Web in 2005 and at subsequent Mars oppositions with the same urgent refrain: "Earth is catching up with Mars [for] the closest approach between the two planets in recorded history. On August 27th . . . Mars will look as large as the full moon." This happened despite the fact that Mars was closest that year in October.

Now dubbed the Mars Hoax, it may well rear its head again when the planet comes to opposition on October 13, 2020; December 8, 2022; ad infinitum. If it does and your friends ask what's up, tell them the story and then direct them to the real thing right outside their windows. Maybe the message will fade away as more and more people realize it's not true. Just about the time it does, Mars will nudge even *closer* to Earth on August 28, 2287, providing another opportunity for a fresh new rumor. Don't even be tempted!

DUST STORMS ON MARS ARE POWERFUL ENOUGH TO DESTROY A MANNED BASE

✦ ✦ ✦

Dust storms are common on Mars. Most behave themselves and remain local or regional. But every so often, a small storm can blow up into a planet-wide monster. In late May 2018, a small disturbance detected by NASA's *Mars Reconnaissance Orbiter* and eagle-eyed amateur astronomers quickly mushroomed. Three weeks later, it became one of the most intense storms ever observed, encircling the entire planet and darkening the skies over NASA's *Opportunity* rover for many weeks.

Loss of sunlight made the rover's solar panels useless, which meant it couldn't recharge its batteries. Without power, the robotic wanderer eventually succumbed to the planet's bitter cold and hasn't been heard from since. NASA officially declared it "dead" on February 13, 2019. It was a shame to lose *Opportunity*, but because it was designed for 90 days and lasted nearly 15 years in the Red Planet's challenging environment, I think we can call it one of the most amazing machines ever built, far exceeding its mission objectives.

Even novice astronomers could tell something was up on Mars during the 2018 storm. Dark markings and the planet's south polar cap, normally visible in smaller telescopes, were shrouded in orange dust and completely obscured.

Mars is dustier than the top of your refrigerator. Orange dust covers much of the surface and drifts through the atmosphere, giving the Martian sky its distinctive butterscotch coloration. The sun warms the ground there just like it does on Earth,

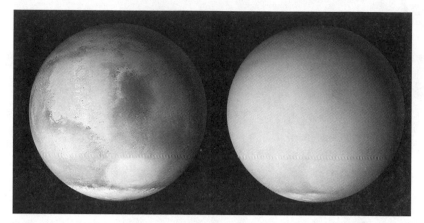

Two 2001 images from the Mars Orbiter Camera on NASA's Mars Global Surveyor orbiter show a dramatic change in the planet's appearance when dust raised by storm activity in the south went global. The images were taken about a month apart. (HST NASA-JPL-MSSS)

heating the air near the surface. Warm air rises, creating winds that carry the dust aloft in a positive feedback loop that can quickly turn a zephyr into a squall.

> "One theory holds that airborne dust particles absorb sunlight and warm the Martian atmosphere in their vicinity," said Phil Christensen, planetary geologist at the University of Arizona, in reference to another global storm in 2001. "Warm pockets of air rush toward colder regions and generate winds. Strong winds lift more dust off the ground, which further heats the atmosphere."

More heat means more energy and stronger winds, which lift even more dust into the air, amplifying a small disturbance into a large one. Occasionally, multiple smaller dust events can coalesce into a bigger regional tempest, driving clouds of powder across vast swaths of the planet. Dust storms are more common during the southern hemisphere summer, when a fair part of the south polar cap rapidly vaporizes in the heat, but storms can occur at almost any time.

When the planet is closer to the sun than usual, as it was in 2018, heating is more intense, which can lead to stronger and longer-lasting storms. At every close approach of Mars to Earth, professional and amateur astronomers carefully monitor the planet for signs of small yellow-orange clouds that could blow up into the next big weather event.

NASA's Mars Opportunity rover stops in Endurance Crater to take a selfie during happier times. The rover struggled through the long and deep dust storm in 2018 until it was pronounced "dead" in Feb. 2019. (NASA / JPL-Caltech)

The Martian atmosphere is more than 100 times thinner than Earth's and is composed mainly of carbon dioxide. A thin atmosphere heats up quickly in the sunshine and cools just as quickly at night. Temperatures can reach 70°F (20°C) on a summer day at the equator but drop to -100°F (-73°C) at night. The average annual temperature on Mars is -81°F (-62°C) compared to Earth's more temperate global average of 53°F (15°C).

Martian winds average about 20 miles an hour (32 kph) and top out at about 60 mph (97 kph). While not tornadic in terms of speed, they're similar to winds produced in severe thunderstorms back here on Earth. This might make you think that the worst Martian winds can really pack a punch, and they certainly did in the movie *The Martian* (2015). The frightening, equipment-toppling gales depicted in the film would certainly give pause to anyone thinking about signing up for Mars astronaut duty. In what was otherwise an excellent movie that got a lot of things right, the storm scene was mostly science fiction, not fact.

One crucial detail was ignored. Because the atmosphere on Mars is less than 1 percent that of Earth, a powerful storm with 60 mph (97 kph) winds would be equivalent to an 8 mph (13 kph) wind on Earth. For an astronaut stuck in the middle of Mars' biggest storm, it would feel like a pleasant summer breeze. Wind chimes would tinkle, but you'd hardly lose your hat, and it certainly wouldn't tear up equipment the way it was depicted in the movie.

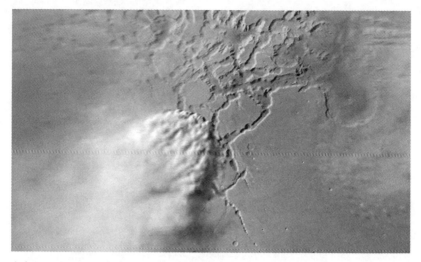

A dust storm encroaches on the Noctis Labyrinthus canyon system on Mars on Nov. 2, 2006. (NASA MRO MSSS)

Drama makes a movie more fun to watch, so it's not surprising that the dust storm became the premise for the plot. That's not to say that Martian dust can't cause trouble. Airborne dust greatly reduces visibility, covers solar panels and insidiously finds its way into equipment, where it can cause electrical and other damage. Tiny in size but many in number, fine particles are guaranteed to pose challenges for future crewed missions to Mars.

Why does Mars have planet-wide dust storms and not the Earth? There are at least two factors involved: the planet's weaker gravity and its lack of oceans. Once atmospheric conditions are ripe for a storm to kindle, wind-borne dust remains aloft longer because of the planet's weaker gravitational pull. Also, Mars has no bodies of water to moisten the air. The added humidity provided by Earth's oceans helps remove dust from the lower atmosphere and slow or prevent dust storms from crossing continents. With no oceans, dust blows hither and yon on the Red Planet, ready to be whipped into a planet-wide pall at a moment's notice.

For all the problems Martian winds can create, they can be helpful too. Dust frequently accumulated on the rover's solar panels and reduced their efficiency. Like the guy that walks out to sweep the stage at the end of a performance, gusty Martian winds have periodically swept the panels clean, keeping them in good working condition and extending the life of both the *Spirit* and *Opportunity* rovers until they, so to speak, bit the dust.

A DOOMSDAY ASTEROID IS ABOUT TO STRIKE EARTH

If you listen to the news or spend time online, it seems like asteroids are whizzing past the Earth all the time. That's a good thing. It means that the several professional sky surveys dedicated to discovering and monitoring new Earth-approaching comets and asteroids are doing their jobs patrolling the skies.

The vast majority of these near-Earth objects, or NEOs, are fragments from asteroid collisions that have been nudged by Jupiter's gravity into orbits that bring them into the inner solar system, where they occasionally pass near the Earth. About 40 new NEOs are discovered each week. Most are small and intrinsically faint; a good number make close flybys of Earth only days after discovery, so they often show up unannounced and depart just as quickly.

If they're big enough and have the potential to strike the planet some time in the future, they're designated as potentially hazardous asteroids, or PHAs. Currently, the Catalina Sky Survey in Arizona and the Panoramic Survey Telescope & Rapid Response System (Pan-STARRS) in Hawaii are responsible for about 90 percent of new NEO discoveries.

As of June 2019, we know of 20,226 near-Earth asteroids (NEAs) ranging in size from 3 feet (1 m) up to about 20 miles (32 km) and approximately 2,000 PHAs; 156 of these are larger than 0.6 mile (1 km). Astronomers define a potentially hazardous object as anything greater than 460 feet (140 m) across that can pass within 4.5 million miles (7.2 million km) of Earth's orbit. It may never collide with the planet, but there's always a small chance in the future that it might.

Asteroid 101955 Bennu, explored by NASA's OSIRIS-REx space mission, has a diameter of 1,614 feet (492 m). It's a potentially hazardous asteroid, one that might someday strike the Earth. (NASA/Goddard/University of Arizona)

Astronomers set the cutoff at 460 feet (140 m) because that was determined to be the minimum size to create a massive *regional* impact if it struck Earth, causing major destruction of a large city and suburbs. If it slammed into the ocean, a devastating tsunami would follow. Impacts of this magnitude occur about once every 10,000 years.

An asteroid several miles across, big enough to destroy the web of human civilization, slams the planet about once every 20 million years. While an event of this magnitude is extremely rare, in the fullness of time, a future impact is all but inevitable. We can only think of the unlucky dinosaurs and their kin who perished in the aftermath of the impact of a 6- to 9-mile (10- to 15-km)-wide asteroid in Mexico's Yucatan about 66 million years ago.

Now for the good news. It's estimated that 93 percent of NEAs larger than 0.6 mile (1 km) have been found, with only about 70 more awaiting discovery. That means we've got our eyes on *most* of the really big threats, and none are predicted to menace Earth as far into the future as we can see. But of the 2,000 PHAs *460 feet* (140 m) or larger, only 20 to 30 percent had been discovered as of 2012. And there are about a million smaller ones in the 130-feet (40-m) range. That's about the size of a thirteen-story building. Only about 1 percent of those have been detected.

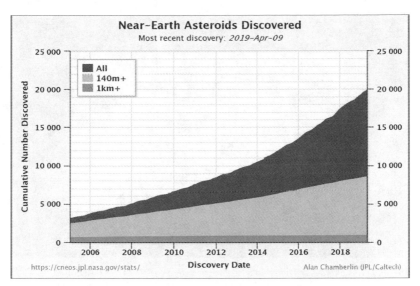

Astronomers have been busy in recent years tallying up an ever increasing number of near-Earth asteroid discoveries. The graphic is current through April 2019 and breaks numbers down in three categories: asteroids one kilometer (0.6 mile) and larger; those 140 meters (459 feet) to 1 km and the total of all discoveries. (Alan Chamberlin [JPL / Caltech])

The tiny asteroid that exploded over Chelyabinsk, Russia, on February 15, 2013, originally measured about 65 feet (19.8 m) wide before it entered Earth's atmosphere, but once it slammed into the air at some 40,000 mph (60,000 kph), it shattered into mostly small pieces. Our atmosphere serves as a great defense against smaller asteroids, pulverizing many of them into dust. Others are fractured into harmless fragments that drop as meteorites.

From half a dozen to about a dozen meteorite falls are witnessed each year. These are the ones you read about in the news, where people see a great fireball and then successfully locate fragments using eyewitness accounts, video security cameras and Doppler weather radar imagery. Based on satellite data and meteor camera networks, it's estimated that about 42,000 meteorites larger than 10 grams (equal to 2½ teaspoons of sugar or two nickels) arrive across the entire planet every year. That's more than 100 gravel-size or larger meteorites a day! Most go unseen over the oceans or arrive in daylight when they're less likely to be noticed. If you add in all the meteorites weighing fewer than 10 grams plus all the dust-size particles that sprinkle the skies, the Earth hauls in between 37,000 and 78,000 tons (33,600 and 70,800 metric tons) of space debris every year.

The dot in this photo is the potentially hazardous asteroid 2014 SC 324. About 165 feet (50 m) in diameter it zoomed only 350,000 miles (570,000 km) from Earth on Oct. 24, 2014. (Gianluca Masi)

There's a widespread misconception that close-approaching asteroids are a danger to the planet. They are if they're headed *straight* for Earth. Thankfully, few are. Like planets, asteroids follow an orbit around the sun and move rapidly, typically at speeds of several tens of thousands of miles an hour. An object moving that fast has so much forward momentum that it can't be pulled in by Earth's gravity. That's why there's no cause for alarm when you hear of an asteroid that will pass closer to the Earth than the moon. Or even closer. In fact, the object that suffers most during a close approach is the asteroid itself. Earth's gravity can alter its orbit, but contrary to popular belief, the planet doesn't "pull" in asteroids.

The closest miss of an asteroid (as of March 2019) occurred on February 4, 2011, when the object 2011 CQ1, estimated at 2.6 to 8.5 feet (0.8 to 2.6 m) in diameter, passed within 3,400 miles (5,500 km) of the Earth and safely departed on its way. On October 13, 1990, the tiny asteroid EN131090, weighing an estimated 97 pounds (44 kg), grazed the Earth's atmosphere and briefly glared as a meteor before returning to space. For a brief time, it was only 61 miles (98 km) above the planet's surface!

About a half million asteroids have names and well-known orbits, so astronomers can predict in advance where they will be. Of course, it's the ones that haven't been discovered that pose the greatest potential hazard. Telescopes on the ground can only spot close-approaching asteroids at night when the sky is clear. Observatory teams use robotic telescopes, which take multiple photos several minutes apart of one region of the sky after another. The images are analyzed for objects that move quickly between exposures, a sure sign that the asteroid or comet is close. Follow-up observations are made to determine an orbit and whether the object poses a hazard.

On three occasions as of this writing, astronomer Richard Kowalski, senior research specialist with the Catalina Sky Survey, detected small asteroids (the meteorite-producing kind) hours before they struck the Earth. The first exploded over Sudan in 2008 and showered the desert with small meteorites, many of which were later recovered by researchers. The second plunked into the Atlantic Ocean in 2014 just 21 hours after discovery, and the third crackled over the skies of Botswana in 2018. All three measured in the 10- to 15-foot (3- to 4.5-m) range. Like the Chelyabinsk or Viñales, Cuba, fireballs, they were headed straight for Earth.

What astronomers presently *can't* observe are all the PHAs that approach our planet in the daytime sky. Invisible in the solar glare, they can sneak up on us. That's why the B612 Foundation, a group dedicated to protecting the Earth from deadly asteroid strikes, tried to fund the building of the Sentinel Space Telescope. From its orbit, the telescope would have been in an ideal position to track the ones that ground-based telescopes miss. Unfortunately, the group didn't meet its funding goals, and is currently assessing less expensive alternatives.

The International Astronomical Union's Minor Planet Center publishes an online list of PHA close approaches to Earth (see Resources on the next page) through the year 2178. Number one on the roster is the ¼-mile (0.37-km) asteroid 99942 Apophis, expected to zing by Earth on April 13, 2029, at a distance of about 24,590 miles (39,600 km). While that means the asteroid will scrunch in closer to us than the satellites that relay TV signals around the world, it will pass safely by. Could it hit a satellite? Not likely. Apophis is a speck compared to the vast amount of space around the Earth in which satellites orbit.

Shortly after Apophis's discovery in June 2004, astronomers gave it a 2.7 percent chance of slamming into the planet in 2029. That's significant. But as frequently happens, more observations yield a better, more accurate orbit. By 2013, the probability of an impact had been totally eliminated.

Asteroid 101955 Bennu, site of NASA's OSIRIS-REx sample return mission (2019 to 2023), has a 1-in-2,700 chance of striking Earth between 2175 and 2199. But due to Bennu's unstable orbit, it's more likely it will fall into the sun sometime in the next 300 million years before it gets a clear shot at us.

When you hear about a recently discovered asteroid that could strike Earth in the next whatever-number-of-years, give it a major grain of salt. As more observations come in, the probability of an impact will almost certainly drop.

I look forward to close passes, new or known, because even intrinsically faint NEOs can become bright enough to see in an ordinary telescope for a few brief hours or days. The closest ones move so fast you can literally watch them slide across the background stars in real time like slow-moving satellites. Every time I observe one of these close shaves I think of the asteroid that led to the demise of the dinosaurs. Hours before it landed, it must have looked just like one of these "moving stars."

Rare as they are, asteroid impacts are inevitable, giving us new respect for these innocent-looking points of light.

Resources

- List of Potentially Hazardous Asteroids: https://www.minorplanetcenter.net/iau/Dangerous.html

- NEO Earth Close Approaches: https://cneos.jpl.nasa.gov/ca

ASTEROIDS ARE THE REMAINS OF A DESTROYED PLANET

+ —+— +

In 1776, German astronomer Johann Titius described a curious relationship between the spacings of the six known planets at the time. He noticed that as you moved outward from Mercury, the innermost planet, each successive planet was roughly twice as far from the sun as the previous.

To find the distances to the planets, à la Titius, begin with the following sequence of numbers: 0, 3, 6, 12, 24, 48. Adding 4 to each and then dividing by 10 gives you each planet's approximate distance from the sun measured in fractions or multiples of the Earth-sun distance, called an *astronomical unit* or *a.u.* The Earth is 1 a.u. from the sun. Do the math and the sequence becomes 0.4 (Mercury), 0.7 (Venus), 1.0 (Earth), 1.6 (???), 2.8 (Mars), 5.2 (Jupiter), 10.0 (Saturn), 19.6 and so on.

J. E. Bode, another German astronomer, formalized the relationship mathematically in 1778, and it came to be known as the Titius-Bode law. When William Herschel discovered Uranus in 1781 at a distance of 19 a.u., it fell right into place, very close to the 19.6 a.u. predicted by the law.

But what about those "???" at 1.6 a.u.? Shouldn't there be a planet there? Well, there sort of is. That distance neatly matches the location of the asteroid belt, which no one knew existed at the time. Herschel's discovery and the seeming vindication of the law prompted Bode to encourage astronomers to look for a "missing planet" in the gap between Mars and Jupiter. Titius agreed, writing: "But should the Lord Architect have left that space empty? Not at all."

18th century German astronomers Johann Bode (left) and Johann Titius, the brains behind the Titius-Bode law. Bode suffered from an eye disease that damaged his right eye. (Public domain)

A group calling themselves the Celestial Police soon formed for the purpose of hunting down the potential new planet. And guess what? They didn't find it. But their heads were in the right place. Italian astronomer Giuseppe Piazzi beat them to the punch, discovering Ceres, the first and largest asteroid, on January 1, 1801. Bode and Titius had nailed it. For a little while anyway. As more and more asteroids were discovered, it became obvious that no true planet existed between Mars and Jupiter. Further, when Neptune and Pluto were added to the planetary lineup in 1846 and 1930, respectively, neither fit into the sequence. Astronomers now see the numerical relationship between some of the planets' distances as more a coincidence than a law of nature.

Asteroid discoveries soon grew from a trickle to a torrent. German astronomer Heinrich Olbers was the first to suggest that the bodies might be fragments of a much larger planet that either exploded or was struck by a comet and shattered to pieces.

That's how we got acquainted with the asteroid belt. As of October 2018, astronomers have discovered 789,069 asteroids, of which they've named 21,787. NASA estimates that the asteroid belt contains between 1.1 and 1.9 million asteroids larger than 0.6 mile (1 km), with a myriad of smaller ones orbiting in a broad band 19.5 million miles (31 million km) wide between Mars and Jupiter.

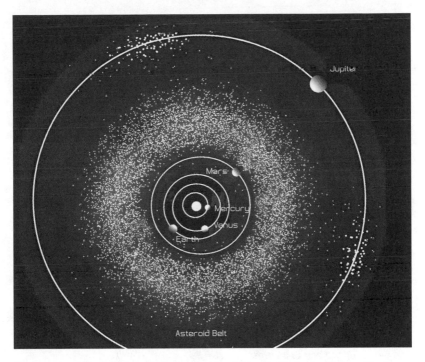

The majority of the inner solar system asteroids orbit in a belt between Mars and Jupiter. Jupiter commands two additional groups of asteroids that orbit ahead and behind it called the Trojans. (NASA)

If we drop the size down to 330 feet (100 m), the number rises to an estimated 150 million asteroids! Only a tiny percentage of these follow Earth-crossing orbits that would make them potentially hazardous as described in the previous section.

We also know that the fragments aren't the remains of a splintered world. Despite the sky-high number of asteroids, if you assembled the entire kit and caboodle into a sphere, the total mass of the asteroid belt would come to just 4 percent that of the moon. Of that, Ceres accounts for a third of the mass of the asteroid belt alone. Hardly enough to build a planet.

But it may not always have been that way.

Several billion years ago, before Jupiter settled into its slot between Mars and Saturn, the bloated gas giant migrated across what is now the asteroid belt, which back then may have held an Earth's worth of mass, plenty to make a planet.

As the solar system's heavyweight Jupiter can't resist stirring things up. (NASA, SwRI, MSSS)

But Jupiter's gravitational energy stirred up the bodies orbiting there, ejecting many from the solar system like a parent clearing a path through a child's toys and depleting the belt to its present state.

Once the belt settled into its current configuration, its troubles weren't over. Jupiter's gravity continued to stir the space rock pot, removing asteroids from the belt or sending them crashing into one another to make smaller asteroids. The pieces never had a chance to come together to form a planet, even a tiny one, because the rate of *destruction* incited by Jupiter's gravitational influence was greater than the rate of *formation*, where smaller bodies could peaceably collect into a larger one through their own self-gravity. Those we see today are true survivors that have managed to skirt Jupiter's meddling since the dawn of the solar system.

Another clue that the space rocks and mountains orbiting in the asteroid belt never formed a planet is found in their diverse compositions, which imply multiple origins. Some asteroids are rich in carbon and water, others in metal and still others are composed mostly of silicate rock like the Earth and moon.

The asteroid belt is a dynamic place where ancient mountains of space rock are still being ground down by meteoroid impacts or pushed by Jupiter's gravity into orbits that can threaten Earth. Instead of a single planet, we have millions of baby ones with endless possibilities.

PLANETS AND STARS CAN FORETELL THE FUTURE

+ ✦ +

Wouldn't we all like the ability to predict the future? Lots of people try. In 1936 the *New York Times* predicted that a rocket would never be able to leave the atmosphere. On the other hand, Nikola Tesla correctly anticipated the cellphone when he predicted in 1909 that "it will soon be possible to transmit wireless messages all over the world so simply that any individual can own and operate his own apparatus." It's always hit-and-miss, but mostly miss by my reckoning.

Astrology is the study of how the positions of cosmic objects, usually stars and planets, determine our personality and other important aspects of our lives. It also makes predictions about the future. Most of us are familiar with astrology through the daily newspaper horoscope, where we can look up our "sign"—a reference to one of the twelve constellations of the zodiac—and read a personal forecast for the day. Astrologers call this *sun-sign* astrology because it's based on what sign or constellation the sun was in the day you were born. People have been following horoscopes since at least 409 B.C.E., the date of the oldest known horoscope chart.

Astrology is practiced in different forms in different countries with no one method shared by all. People who are really into it will tell you that the newspaper horoscopes are too simplistic and don't take into account the planets and their influences. For a more complete picture you'll need a full reading based on the date, time and location of your birth.

Thousands of years ago, no one knew about the *physical* nature of stars and planets. Planets were gods, each attached to his or her own crystalline sphere that spun around the immobile Earth. The stars occupied the outermost sphere

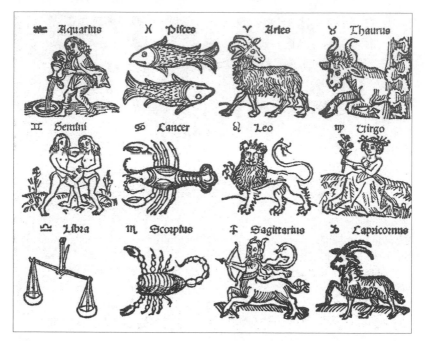

A 16th-century woodcut of the 12 zodiacal signs. (Public domain)

and were made of ether—a "fifth element" after earth, water, wind and fire. The stars and planets were believed to be eternal, and unchanging compared to the Earth—a hot mess of four changeable elements. Earth was different, made of the four changeable elements only.

Whether or not they got the details correct, our distant ancestors sought a connection with the sky through its twinkling lights. Light pollution was nonexistent in those days. When the moon didn't interfere, star-studded skies were the rule.

Those early cosmic connections were both practical and spiritual. People used the stars to mark the best times for hunting and planting and for determining the dates of important religious festivals. Much of western civilization has severed these connections save for the start of the seasons, which are still determined by the sun's position in the sky on the dates of the vernal and autumnal equinoxes and summer and winter solstices.

Sagittarius "Teapot"

Capricornus

Jupiter

Saturn

Mars

Facing Southeast at Dawn on April 9, 2020

A beautiful planetary alignment at dawn on April 9, 2020. Gatherings of the planets are a common occurrence as they periodically come together along the same line of sight though at vastly different distances from us. (Stellarium)

Many of us still look to the stars to anticipate the march of the seasons. When we see Orion return to the morning sky in September, we're reminded that winter will soon be on its way. Saturn takes 29.5 years to go once around the sun. I first identified the planet in the constellation Aquarius when I was just eleven years old. The next time it returned to that spot I was married with two children and in the middle of my career. Saturn's cycle had become a touchstone for me.

Like we still do to this day, our ancient ancestors sought a spiritual relationship to the cosmos. Astrology helped make that connection because it related people's daily lives to the realm of the gods and the grand cycles of the stars and planets.

I used to look at my horoscope for hints of what might be coming down the road. Maybe there was a nugget in there for me. Then one day, I decided to read *all* the signs, not just my own. That's when I discovered that the character descriptions and advice for many of the other signs fit as well or better than my own. I'm officially a Leo, but I also seem to be a Scorpio, a Virgo, a Gemini, etc. I realized that it was only wishful thinking that made my horoscope appear customized for my life. That's when I stopped paying attention to horoscopes. My fall from astrology had begun.

I could argue that there's no physical connection between stars and planets and you and me. As we learned in the earlier section on comets, the whole lot of solar system bodies outside of the sun and the moon have next to no gravitational effect on the Earth and its inhabitants. The same is true for the stars. Being so remote, their influence is even less. Nor does a planet's position on the date of your birth or the angles it makes with other planets or even what "house" it's in have any measurable effect on your life. Not physically, anyway. It was and still is all magic to me, based on wishful thinking and confirmation bias, the human tendency to remember the "hits," when an apparent prediction proved correct, and forget the misses, when it didn't.

A separate branch of predictive astrology uses planet transits and solar and lunar positions to forecast future events. In the days of the gods, we might convince ourselves that their powers could do such things, but in the twenty-first century we know that celestial bodies are fascinating real places you can actually travel to, blissfully unaware of their power to influence human affairs. Do they really conspire to alter the future? Predict our next relationship?

If I sound critical of astrology, I am. It's not based in science, and there's no experimental evidence to substantiate its truth. While Saturn's cycles may be an important marker in my life, I don't believe Saturn's movements affect me personally.

Still, I'd be the last person to call us creatures of logic. I need to look no further than my splendidly imperfect self. Bottom line, we want to be spiritually connected to the universe, feel a small part of something much greater, be it God, nature or the spiritual forces behind astrology. Or all of these. Desire for that deep feeling sharpens and renews our spirit and makes us better people.

I can only speak for myself. Scientific discovery, the human spirit and nature provide my daily dosage of vitamin W (wonder), but if astrology orders your world and satisfies your need for a spiritual connection, I won't call in the Spanish Inquisition. We're both humans just trying to find our way.

Precession

For the record, you'll sometimes hear that the signs used in astrology are off by a full constellation. All signs were assigned several thousand years ago during astrology's formative years. On the date of your birth today, the sun is removed one constellation to the west of its official "sign." I'm a Leo—or was. At my birth, the sun shone from Cancer the Crab and has on every birthday since. The westward shift of the sun's position through the zodiac constellations is caused by a wobble of the Earth's axis called *precession*, discovered around 150 B.C.E. by the Greek astronomer Hipparchus. Briefly, precession has shifted the sun's position in the zodiac 1.4° per century. Over 2,000 years, that adds up to 28° or about one constellation width.

Astrologers recognize precession and have fixed the twelve zodiac signs to twelve identically wide swaths of sky along the sun's path from a reference point known as the First Point in Aries, the location of the sun at the spring equinox roughly 2,000 years ago. Today, your *sign* will be the same as it was decreed long ago, but the location of the sun on your birth date has shifted one constellation to the west in the current era. I could argue that astrologers should take precession into account . . . but I won't.

JUPITER IS A FAILED STAR

+ ✦ +

Blame it on the sun's sticky fingers. It clawed the lion's share of the matter in the original solar nebula to itself, leaving too little for Jupiter or any other planet to grow massive enough to become a companion star. Instead, Jupiter settled for the title of "biggest planet," with a mass 318 times that of Earth. Big enough that if you hollowed it out like an oversize pumpkin, you could squeeze nearly 1,300 Earths inside.

As huge as Jupiter is, size isn't so much the issue as mass. Even though Jupiter is primarily made of hydrogen and helium like the sun, it simply doesn't have enough stuff on board for gravity to crush and heat its core to the tens of millions of degrees required to fuse hydrogen into helium. That single skill is the most basic requirement for a glob of matter looking to become a star. In true stars, four hydrogen atoms ultimately come together in the fiercely hot and pressurized environment of a star's core, fuse to form helium and in the process convert 0.71 percent of their original mass into pure energy, along with a blast of lightweight neutral particles called *neutrinos*.

When it comes to fusion, you get a lot of bang for your matter buck because everything from hamburgers to hazelnuts is little more than a super-condensed form of energy. Einstein proved it with his most famous equation, $E = mc^2$. "E" stands for energy, "m" for mass and c^2 is the speed of light squared (times itself). To find out how much energy a bit of matter contains, you only need to multiply its mass (m) times the speed of light squared. Try it once and you'll end up with a colossal number, the reason why even itty bits of matter—including that used to make a thermonuclear bomb—contain phenomenal amounts of energy.

Jupiter is the solar system's most massive by far. Yet it would have to be 80 times weightier to transform itself into a second sun. (Lsmpascal CC BY-SA 3.0)

Diagram showing the relative sizes of Jupiter, brown and red dwarf stars and the sun. Brown dwarfs aren't true stars; red dwarfs are. (NASA / Goddard)

One gram of water converted to pure energy would produce an explosion equal to 20,000 tons (18,143 metric tons) of TNT. Deep inside the sun's core, temperatures sizzle at 27 million degrees Fahrenheit (15 million Celsius) under pressures more than 340 *billion* times that at sea level here on Earth. Energy released by fusion begins as powerful gamma rays but gradually loses energy as it bounces off a dense soup of subatomic particles on its way to the surface. By the time radiation arrives there, it has traveled some 400,000 miles (644,000 km) in distance and 1 million years in time. It's lost so much energy through collisions that it departs the sun primarily as visible, infrared (heat) and ultraviolet light. And it really does take that long to reach the surface. Every sunbeam tanning your back is older than the hills.

If you could take about 80 Jupiters and cram them all together into a ball, you'd have your star. Pressures and temperatures in its core would be enough to initiate and sustain fusion. You might imagine that an 80-mass planet would be enormous, but that wouldn't be so. Because the force of gravity increases with increasing mass, gravity would scrunch everything down into a sphere only modestly larger than Jupiter is today. For perspective, the sun is about 1,000 times more massive than Jupiter but only about 10 times its diameter.

The smallest true stars are the red dwarfs with a minimum mass of 7.5 percent that of the sun and 78 times that of Jupiter. Enough to put a "fire" in a star's belly, so to speak. Proxima Centauri, a faint member of the Alpha Centauri system, the closest star system after the sun, is a typical red dwarf with 12 percent the mass of the sun and a diameter of about 125,000 miles (200,000 km) or 1.4 times the size of Jupiter.

Stars form in essentially one step, from the direct collapse of interstellar clouds of gas and dust. Leftover materials orbit around the newborn sun in a flattened disk. Planets assemble from the leftovers in a two-step process. Smaller chunks of ice and rock coalesce first, and these act as embryos that gather more material through gravity to ultimately make a planet. Jupiter followed the planetary path, while Proxima Centauri and its ilk took the more direct route to "stardom."

You may have heard about another type of star called a *brown dwarf*. These are the true "failed stars." Brown dwarfs are about the same size as Jupiter but 15 to 75 times as massive, with cores warm enough to fuse small amounts of lithium and deuterium (an alternate form of hydrogen called an *isotope*) but lacking the heat

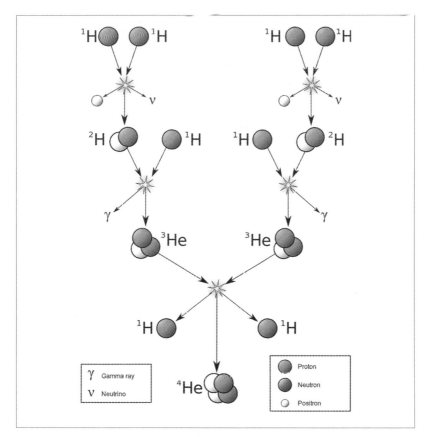

When a body is massive enough for hydrogen atoms (H) to fuse to form helium (He) in its heated, compressed core, it ignites as a star. Energy released as gamma rays in the process makes it self-radiant. (Public domain)

and pressure to sustain nuclear fusion like true stars. Some of them are even cool enough for methane, a gas also detected at Jupiter and the other outer planets, to form in their outer atmospheres.

I wouldn't call Jupiter a failed anything. Because of its gravitational might and position, it has tamed many a comet, bending their paths to its will, even flinging them out of the solar system altogether if they stray too close. Today, it still routinely nudges asteroids into orbits where they can pose a threat to Earth in the distant future. Star or no, Jupiter rules within the planetary domain.

THE NEMESIS DEATH STAR EXISTS AND CAUSES MASS EXTINCTIONS

✦ ✦ ✦

It's a pity the sun missed out on being a double star. As a single sun, it's something of a rarity. Astronomers estimate that up to 85 percent of stars have companions. There are so many double stars up there, you can pick any random fist-worth of sky, point a telescope there and rack up at least a few pairs. Some like Albireo in the Northern Cross are among the loveliest sights in the sky. Others are set so close together you can spend a half hour at different magnifications trying to separate them . . . and love the challenge.

We may never know whether the sun had a companion, but a May 2017 study by a joint U.S.–German team of astronomers predicted that all stars may initially form as pairs or multiples separated from each other by 500 times the Earth–sun distance. Some gradually fall toward one another into tighter orbits, while others break up into separate stars. Who knows? Four and a half billion years ago, two suns may have blazed at the center of the newborn solar system and cast twin shadows across the newly formed planets like the famous twin-suns scene on the planet Tatooine in *Star Wars*.

In the early 1980s, scientists observed that mass extinctions on Earth occurred about every 26 million years. Richard Muller of the University of California, Berkeley speculated that the cause could be periodic comet storms from the Oort Cloud, a repository of billions of icy comets orbiting at the fringe of the solar system. To get the ball rolling, he suggested that the sun had a companion, a red dwarf star 1.5 light-years away in a long, cigar-shaped orbit. Every 26 million years, the star

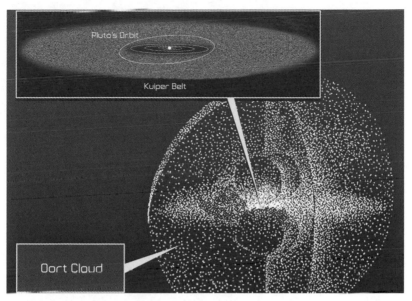

The solar system extends far beyond Pluto to the Kuiper Belt, home to icy asteroids and comets, all the way to the remote Oort Cloud, a vast comet repository 2,000 to 100,000 times the Earth-sun distance away. The hypothetical planet Nemesis was thought to cause periodic mass extinctions by tossing comets from the Oort Cloud toward the Earth. (NASA)

would pass close to the Oort Cloud, shake a bevy of comets loose with its gravity and send them sailing toward the sun to wreak havoc on the inner planets.

Based on distance and type, Muller estimated the dwarf's brightness as between 7th and 12th magnitude, bright enough to see in a modest-size telescope. To find it, astronomers would need to examine each red dwarf in the sun's neighborhood in search of one that moved rapidly compared to more distant dwarfs. Muller called his speculative star Nemesis, a fitting name for a species killer.

The idea had merit, and a number of scientists considered the idea plausible, but despite searches and surveys, no trace of Nemesis was found. A search by the University of California Leuschner Observatory in the early 1980s turned up no candidates. All-sky surveys by the Infrared Astronomical Satellite (1980s), the 2MASS Survey (1997 to 2001) and NASA's WISE mission (2009 to the present day) also turned up nothing.

Surveys done in infrared light are the most telling because cooler stars like red dwarfs and even pseudo-stellar brown dwarfs shine most brightly in infrared. We

The brilliant white dot that looks like a putative planet or "second sun" is nothing more than an internal reflection in my mobile phone's camera. (Bob King)

should have found Nemesis by now if it were as close as advertised. But at least for the moment, the sun remains a loner.

When new information or ways of looking at information are discovered, scientists will often re-examine their previous analyses and assumptions. A more recent mathematical analysis of the extinction events no longer shows the 26- to 27-million-year pattern found in earlier analyses. In fact, there are no regular intervals between major extinction events, only iffy ones every 62 million and 140 million years.

Though no longer needed to explain extinctions, Nemesis would not die. To this day it's occasionally conflated with Nibiru (see page 118) or with bright spots near the sun in pictures taken with mobile phones. Go to YouTube (or similar social media) and you'll find earnest videos taken by folks holding their phone up to the sun and describing a "second sun" in the same field of view. Sure enough, bright spots are clearly visible, but we needn't invoke Nemesis to explain them. What the clips show are internal reflections of sunlight within and between the camera's lenses. I've been seeing—and trying to avoid—these bright spots for decades. One way to know whether you've got an internal reflection rather than a second sun is to put the phone down, block the sun with one hand and try to see it with your eyes. If only your phone shows it, it's not real.

Sundogs or parhelia form on either side of the sun when sunlight is refracted through ice crystals. Occasionally, they appear nearly as brilliant as the sun itself. (Bob King)

Other round objects sometimes appear near the sun and grow intensely bright, sometimes rivaling it in brilliance. These are real and called *sundogs*. They usually come in pairs (though they can be single, too) and appear about two outstretched fists (22°) on either side of the sun. They're a meteorological phenomenon that happens when ice crystals in high clouds refract sunlight into symmetrical patches flanking either side of the sun. I've seen sundogs bright enough to make me stop the car and jump out to take a photo. Rest assured, these "suns" are momentary and much closer to home, just 6 to 7 miles (10 to 11 km) up in the atmosphere.

The more you engage with nature, the less likely you are to jump to conclusions when confronted with something unfamiliar. Regular observation of the sky provides a full-course meal for your curiosity. And once you see and identify a phenomenon, you'll become "sensitized" and likely see it again. When I'm looking for mushrooms in the woods, another hobby I enjoy, I might get lucky and spot a chanterelle, one of the tastiest of fungi. I'll examine the mushroom and its immediate environment carefully to train my eye to know what to look for. That way I'm more likely to find other chanterelles that might be nearby.

Paying attention is a skill worth cultivating every day of your life.

THE PLANETS ORBIT THE SUN IN CIRCLES

+ + +

Diagrams often depict the planets orbiting the sun in circles. But no planet we know of follows this perfect path. Nature prefers a more relaxed version of the circle called an *ellipse*. All the planets revolve around the sun in *elliptical* orbits, and all the known moons—nearly 200 as of last count—obediently trace ellipses about their host planets. Comets and asteroids? More ellipses.

An ellipse has an oval shape that resembles a pill you might find in your medicine cabinet. Or a squashed circle. You can make an ellipse by sticking two pushpins 1 inch (2.5 cm) apart into a piece of cardboard. Next, cut a length of string about 6 inches (15 cm) long and tie it in a loop. Place it over the pushpins and hook a pencil tip into the string. Pull the string tight around the pins, and you'll sketch out an ellipse.

The two pins are called the *focal points* of the ellipse. The closer together they are, the closer the ellipse approaches a circle. When the focal points exactly coincide (or if you just use one pushpin instead of two), you'll trace a circle, which is considered a special type of ellipse. If you separate them farther and use a longer string, your ellipse will be thinner, shaped more like a cigar.

The more "squashed" an ellipse is, the larger its *eccentricity*. Eccentricities vary from 0 to just under 1, with 0.0 being a circle and 0.999999999-etc. a long, thin ellipse resembling a dinner plate viewed from the side. Earth's orbit is nearly a circle with an eccentricity of just 0.0167.

In a planetary orbit, the sun sits at one focus of the ellipse. The other focus is empty and plays no role in the orbit. Let's take Earth as an example. As the planet moves along its orbit, it comes closer to the sun at one end of the ellipse and farther away at the other end. When it's closer, it feels the strength of the

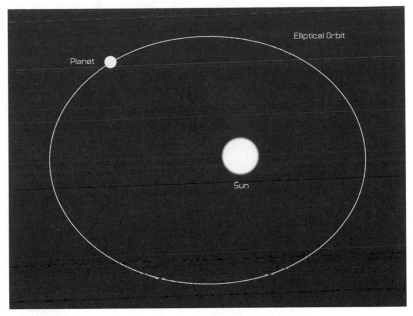

Planets orbit the sun in a "relaxed" circle called an ellipse. (Bob King)

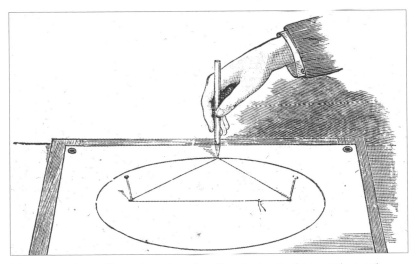

It's easy to make an ellipse using a piece of cardboard, a length of string and a pencil. (Public domain)

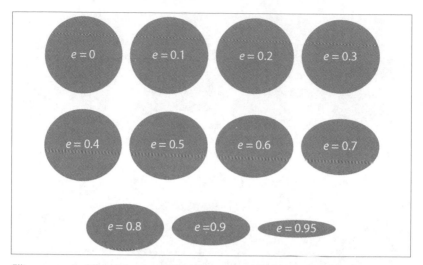

Ellipses come in different degrees of flatness or eccentricity. Planet orbits have low eccentricities close to circular, while distant comets often have more eccentric, cigar-shaped orbits. (Amit6 CC BY-SA 3.0)

sun's gravity more strongly and moves a little faster. When farther away, it slows down. Over the course of its yearlong orbit, Earth's speed varies by more than 2,000 mph (3,200 kph) and its distance by 3 million miles (4.8 million km).

Mercury's orbit is the most squashed, with an eccentricity of 0.21; its distance from the sun varies dramatically around its orbit, from 29 million to 43 million miles (46 million to 70 million km). Despite its relatively large eccentricity, if we could see Mercury's orbit from far above, it would still look pretty close to a circle. You only start to notice a circle "going flat" when the eccentricity reaches about 0.3.

Circles seem simple, so why don't planets have circular orbits? Like so many things, a circular orbit *looks* easy when, in truth, it's exceedingly difficult to maintain. If you had superhuman powers and could start a planet off in a perfectly circular orbit with a constant velocity around the sun, it wouldn't be long before it evolved into an ellipse. Nothing exists in total isolation. The gravities of the other planets (especially Jupiter) would tug this way and that on your planet, altering its speed just enough to convert its once-circular path into an elliptical one.

Because Mars is quite a bit closer to Jupiter than the Earth is, Jupiter's gravity changes the eccentricity of Mars' orbit and the tilt of its axis. Both have been and still are significant factors in altering the planet's climate over time. In the early days of the solar system, when the planets' orbits and locations were sorting themselves out, some of them may have been whizzing around the sun in highly eccentric orbits. Interactions with Jupiter and the other planets may have sent these wayward worlds careening off to the stars. The current planets likely survived the chaos because they settled into nearly circular orbits early on and stayed out of each other's way. They played nice.

Everyone from the ancient Greeks to Copernicus, the Polish astronomer who proved we lived in a sun-centered solar system, got the orbits wrong. You can blame it on cherished beliefs. The heavens were always seen as a place of circular motion based on the belief that circles and spheres were the most perfect geometrical shapes. Yet for all the things Copernicus got right, his model of the solar system couldn't predict the movements of the planets with precision because he was stuck on circles. It wasn't until the seventeenth century, when German astronomer Johannes Kepler discovered that the planets revolved around the sun in *ellipses*, that planetary motion started to make sense. Think how many thousands of years we were beholden to the notion of perfect circles. It makes you wonder how many other things we take for granted that may not be as they seem.

SUN, STARS AND SPACE

YOU CAN TRAVEL FASTER THAN THE SPEED OF LIGHT

+ ✦ +

"Full ahead, Mr. Sulu, maximum warp." Whether it was rescuing the crew of a stranded ship or outrunning the Klingons, the crew of the *Star Trek* series could count on their warp drive to get them out of trouble. Traveling at the speed of light or better, it made for both quick escapes and timely rescues. Warp 1 in the original series corresponded to the speed of light, or 186,000 miles per second (300,000 km/sec). Warp 2 equaled 8 times the speed of light (2 x 2 x 2) and Warp 9—the ultimate speed limit for the *Enterprise*—would shoot the crew across the universe at 729 (9 x 9 x 9) times faster than light speed. If humans really could travel this way, I'd just warp to the Orion Nebula instead of freezing my fingers looking at it through my telescope in January.

The concept of warp drives first surfaced in sci-fi books several decades before the TV program. Now they're a mainstay in movies, books and videos when it comes to tackling the headache of cosmic distances. Without this little plot device, can you imagine how boring these shows would be? Even traveling at the incredible but still feasible speed of 99.1 percent light speed, it would take almost seven months to get to Alpha Centauri, the closest star after the sun. Yawn.

In a wonderful twist, warp drives work by warping space to shrink the distance to your destination, not by propelling the ship itself at faster-than-light speeds. Think of space as a pair of pants and imagine yourself the size of an ant. As a teeny-weeny person you could take the long way and walk from the waistband down to one of the cuffs. But if instead you *folded* the pants in half, pinching the waist and cuffs together, you'd drastically reduce your travel time. *Voilà*, folded space!

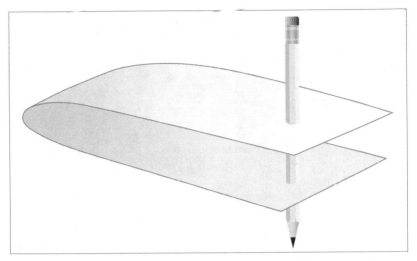

If you could fold space you might drastically reduce the time it would take to travel to distant regions of the universe. (Public domain with additions by Gary Meader)

If your warp drive is balking because you're low on dilithium crystals, your next best bet is a wormhole. Wormholes, which may or may not exist, and if they do, may or may not be navigable, are another favorite ploy for getting from one place to another in a hurry. You enter one end of what looks like a Chinese finger trap and pop out at the other, hoping the Klingons in their haste missed the turnoff.

Is any of this possible? Maybe. While warp drives are a *long* way off, wormholes are theoretically possible, even predicted by general relativity theory. In the 1930s, Einstein and physicist Nathan Rosen worked on the concept of "bridges," or shortcuts connecting different regions of space-time that we now call Einstein-Rosen bridges. Even if a wormhole were large and stable enough to allow safe passage (they tend to collapse without warning), and humans could be protected from the exotic matter of which they're thought to be made, we've yet to find any evidence of them.

One of relativity theory's central tenets is that the speed of light is the universal speed limit of *matter*. This means that a ship or anything else made of matter could approach the speed of light but never achieve it. Space has no such restriction. Space, as in the space between galaxies in our expanding universe, can balloon at faster-than-light speeds without breaking a sweat.

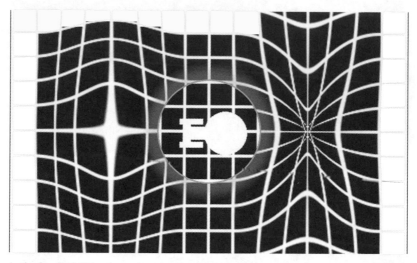

In the Star Trek TV series and movies, the ship leaped across great distances in a matter of minutes by warping space to shrink the distances between destinations. (Public domain / Wikipedia)

Einstein predicted, and experiments have shown, that the faster something travels the more massive it gets, and the slower time flows from the perspective of an outside observer. As the speed of the object increases, the more energy it takes to move it. You and I would never notice these changes at "daily life" speeds like freeway driving or flying in a plane. Only as you approach light speed do these bizarre effects begin to show. Scientists have performed experiments speeding up electrons, the tiny particles that spin around the nucleus of an atom, and confirmed Einstein's prediction: the faster the electrons moved, the heavier they became and the more energy it took to increase their speed.

If you could make something even as minute as an electron travel at the speed of light, it would become infinitely massive and possess an infinite amount of energy. As well, time would slow to a stop for that electron from our point of view. We would see it stop moving at the same time it plugged up the entire universe. Ugh! Of course, this is impossible and could never happen because—nod to Einstein again—matter can't travel at the speed of light.

Joseph C. Hafele (left), a physicist, and astronomer Richard E. Keating monitor two atomic clocks set up on a commercial airliner while performing the around-the-world clock experiment in October 1971. (AP photo)

What does the electron see as its speed ramps up? From its perspective, time would tick by normally. The particle would only notice that it took much less time to get places. Instead of 10 nanoseconds to reach the edge of its container, only 5 nanoseconds would elapse. From its perspective, the distance—or space—between things would have shrunk.

If the electron looked back at the experimenter, she would appear to move in slow motion and her clock tick more slowly. These two different perspectives—of time flow—define the essence of relativity in relativity theory. Einstein discovered that space and time were relative, not absolutes—the flow of time and distance depends on one's frame of reference. The only absolute is the speed of light.

Time dilation, the difference in what clocks read depending on their speed, is as real as that electron putting on weight. Physicists Joseph Hafele and Richard Keating flew four extremely accurate, synchronized atomic clocks on commercial airplane flights in 1971. The planes flew twice around the world, first east and then west. When the times were compared with the one back in the lab at the U.S. Naval Observatory, the traveling ones ran slower by exactly the amount predicted by Einstein.

The only thing we know of that travels at light speed is light itself. Light can behave as both a wave and a particle. Particles of light are called *photons*. Photons have zero mass, and because they're moving at the speed of light, they experience neither time nor distance. If you thought distances shrink for a fast electron, they REALLY contract if you travel at light speed. Photons arrive at your destination *instantaneously*. Meanwhile, an experimenter on Earth looking at your clock would say that it had stopped. Because a photon can get anywhere in *no* time it's everywhere in the universe simultaneously. Distance doesn't exist! I feel a brain explosion coming on.

If all goes according to plan, NASA's *Parker Solar Probe* will become the fastest human-made object ever in December 2024, when it reaches its closest point to the sun, traveling at 430,000 mph (632,000 kph). That's equal to 0.064 percent the speed of light. Though impressive, it doesn't hold a candle to a beam of light. For now, we'll have to content ourselves with movies and TV, where you can travel halfway across the galaxy in just an hour, including commercial breaks.

Earlier we touched on space breaking the light barrier. Can anything else? The short answer is no. Some phenomena can mimic faster-than-light motion. Light-emitting jets of material blasting from supermassive black holes can appear as if they're moving toward us at superluminal speeds. But it's only an illusion caused by the material moving at nearly the speed of light to begin with. As the end of the jet (the part closest to us) gives off light, the light leaving the section close behind is right on its heels, making it look like a stream of light coming from the jet is moving faster than light.

If the thought of attaining light speed in the real world has you bummed, just remember that the planet we live on is humming around the sun at 18.5 miles per second (30 kms), and the sun in turn is traveling at 486,000 mph (782,000 kph) around the center of the Milky Way Galaxy. No matter what, you're still moving faster than any spaceship on Earth can take you. I hope you're enjoying the ride as much as I am.

Resources
* The Theory of Relativity—An Easy Guide: curious.astro.cornell.edu/physics/the-theory-of-relativity

THE NORTH STAR IS THE BRIGHTEST STAR

When I was younger, I'll never forget the surprise in people's voices when I showed them how to find the North Star using the two Pointer Stars at the end of the bucket of the Big Dipper. "That's IT?" someone would ask. My friends and neighbors expected a much brighter object. "Yep, that's it," I'd say, "46th brightest."

Also known as Polaris, the North Star is hardly a slacker. At magnitude 2.0, it holds its own against the brighter stars in the Big Dipper. That makes it easy to see even from the suburbs. What makes Polaris famous isn't radiance but position. It's the closest bright star to the north celestial pole, the point in the sky directly above Earth's north polar axis. Never moving from its appointed spot, it makes the perfect direction indicator. Generations of people have counted on Polaris to point them north.

If you could spend a clear night at the North Pole, the North Star would shine from directly overhead. You could then picture Earth's axis as shooting straight through your body and upward in the direction of the star, located 445 light-years away. As you travel south from the northern hemisphere along the curve of the Earth, the angle between you and the north polar axis increases, causing the North Star to dip ever closer to the northern horizon.

Drop down to a lower latitude—say Winnipeg, Manitoba, at 50° north—and Polaris would now stand 50° (five fists) high in the northern sky. Continue to Honolulu at 21° north and the star would twinkle between the palm fronds only 21° or two fists high. At the equator, Polaris stands level with the northern horizon. Below the equator it dips below the northern horizon and disappears from view. The height of the North Star above the horizon is equal to the observer's latitude.

No matter the season, if you can find the Big Dipper you can find the North Star. Just use the two stars called "The Pointers" to point you there. (Stellarium)

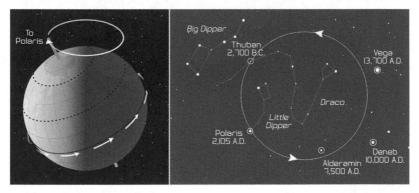

The Earth's axis slowly wobbles or precesses, describing a complete circle in the sky every 25,800 years. Stars lying along or near the circle take turns as the polestar. Polaris in the Little Dipper is our current polestar, but in about 11,000 years Vega will assume that role. (NASA [left], Stellarium)

When the Great Pyramid of Giza was built Thuban in Draco the dragon stood at the pole. (Stellarium)

One thing doesn't change in all your travels. Polaris will always be found at the north polar point in the sky. No wonder it's been so helpful for finding directions at night. If you face Polaris, you face north. Extend your arms out on either side of your body, and the right arm points east, the left west and your back faces south.

OK, I fibbed a little. The North Star really does move, but our lives are too short to notice. Earth's rotation causes our planet's equator to bulge outward 26.5 miles (43 km) like a middle-age "tire" around its rocky belly. The sun and moon tug on the bulge, causing the planet to gyrate on its axis in a motion called *precession*. Like a top that spins off kilter, the Earth's axis traces a circle in the sky while still maintaining its 23.5° tilt. Since the direction the axis points determines which star will assume the role of pole star, that can only mean one thing: Polaris hasn't been the North Star forever.

In fact, around the year 1,000 C.E. it stood 7° from the pole. If you kept track of Polaris's position during the night back then, you'd notice it described a small circle around the north celestial pole wider than your fist. While not nearly as close as it is now, Polaris was still the closest bright star to the pole at that time and useful as a direction indicator. Let's go back further to the year 1 B.C.E., when

Polaris stood 12° from the pole, further than Kochab, the next brightest star in the Little Dipper. People then found north by looking roughly midway between the two stars at a relatively blank spot of sky.

In 2,560 B.C.E., when Pharaoh Khufu looked over what would become his future tomb, the Great Pyramid of Giza, the star Thuban in the tail of Draco the dragon occupied the northernmost spot in the sky, just 1° from the celestial pole.

One full precession cycle takes about 26,000 years. Polaris will return to the same spot 26,000 years from now, around 28,000 C.E. Between times, our descendants will see a slow march of polestars that includes Deneb in the Northern Cross, Kochab and brilliant Vega. Vega will punch the precessional clock about 11,000 years.

In our era, Polaris has been a worthy pole star, currently just 0.6° from the pole, a little more than the apparent diameter of the full moon. And it's still closing. Closest approach happens around the year 2105 when the star slides 0.5° from the due north position. Then it slowly slides away in the coming centuries and won't be replaced by a reasonably bright naked-eye star until around the year 3500, when third-magnitude star Gamma Cephei steps up to the task.

Enjoy the current keeper of the pole. The next clear night, give it a long stare and try to appreciate that you're looking at a star 2,000 times brighter than the sun and 45 times as wide. If you plucked it from its perch and set it in the center of the solar system where the sun is now, Polaris would appear more than 22° across from Earth, big enough to cover the constellation of Orion. Bright or not, it's a big deal.

THE SUN WILL ONE DAY EXPLODE AS A NOVA

+ ✦ +

It better not! Thank goodness we don't have to worry about this ever happening. The future sun may have bad things in store for the Earth, but at least it won't blow up. We orbit a yellow dwarf star that's reliably fusing hydrogen into helium in its core, creating the energy that makes Earth the best place to live in the universe—that we know of. Like cords of wood stacked in the shack around back, the sun's hydrogen will keep its furnace running for a long, long time.

Born from a collapsing cloud of dust and gas about 4.6 billion years ago, the sun reached early middle age just about the same time humans began to understand its true nature. It has another 4 billion years of good living before changes in its core spell the end of the star as we know it. But I'm getting ahead of myself.

The word *nova* means "new" in Latin. In astronomy, it refers to an explosion on the surface of a white dwarf, a star only the size of the Earth but so incredibly dense that a teaspoon of it weighs 15 tons (13.6 metric tons). A nova occurs in a close binary star system where one member is a white dwarf. The white dwarf pulls hydrogen gas from its companion into a stream, which spirals down to the dwarf's surface and piles up. Compressed by the star's intensely strong gravity, it reaches a critical temperature and density and ignites in a massive and brilliant thermonuclear explosion.

In a matter of hours, the star becomes 50,000 to 100,000 times brighter than the sun. Amateur observers who patrol the sky for novae are often the first to record the "new" star. Of course, the star isn't really new. It's been there all along but too faint to draw attention.

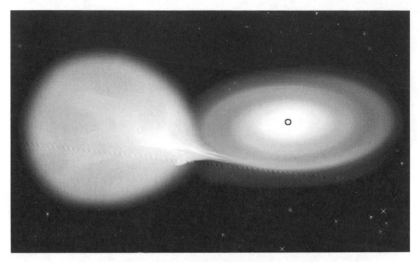

A nova happens only in a very close binary star system where a white dwarf draws matter from its companion. The material spirals down to the dwarf's surface until enough accumulates to ignite in a thermonuclear explosion. (NASA-CXC-M.Weiss)

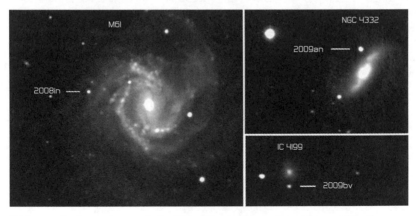

We're overdue for a supernova in the Milky Way, but hundreds are discovered in other galaxies every year. They're typically visible for months at a time and look like faint stars. These are a few examples from 2008–2009. (William Wiethoff)

Surprisingly, both stars survive the blast, and the dwarf resumes its pilfering, resetting the timer for another explosion thousands of years in the future. All novae involve two stars, one to supply the fuel and a white dwarf to steal it. Because the sun is single, we can rule out the possibility that it will ever "go nova."

What about a supernova? Professional surveys and amateur astronomers discover hundreds of supernova each year, virtually all of them in faraway galaxies. Supernovae occur when a white dwarf star burns and explodes or when a supergiant star like Betelgeuse in Orion runs out of fuel and implodes. Let's examine each in more detail.

In the former, we return to the white dwarf–companion star scenario, but this time the dwarf siphons too much material from its mate. If more than 1.4 times the mass of the sun accumulates on its surface, the dwarf can no longer support its own weight. A catastrophic collapse follows, converting the entire star into an enormous thermonuclear bomb. Kablooey!

The explosion is so titanic, even smaller telescopes (6 inches [15 cm] and up) can see the brightest supernovae in galaxies tens of millions of light-years away. Mergers with other stars can also send the dwarf over the 1.4 solar mass limit and initiate a supernova conflagration. This tipping point is better known as the Chandrasekhar Limit because it was discovered by the brilliant Indian astrophysicist Subrahmanyan Chandrasekhar earlier last century.

Supergiant stars at least eight times more massive than the sun can also become supernovae. Isn't it fascinating that both the tiniest and the most massive suns are prone to violent ends? Huge stars are victims of their ability to manufacture ever more complicated elements in their cores. Hydrogen burns to make helium, which fuses to make carbon and oxygen. Carbon and oxygen fuse to form neon, which fuses with other elements to make magnesium. Oxygen burns to silicon and silicon fuses to make sulfur, calcium, nickel and finally *iron*.

The massive star began like the sun, burning the simplest element, hydrogen, into helium, then cooked helium into ever more complex elements. By the time a supergiant star runs out of fuel, the core looks like the inside of an onion: hydrogen burning in the outermost layer, helium in the next level down, carbon below that and so on all the way until we arrive at the iron core.

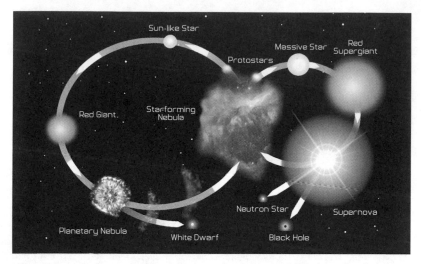

When it comes to stars, everything begins as a cloud of gas and dust (center) called a nebula. Depending on the cloud's mass it can collapse to form a small, sun-like star (left loop) or a massive supergiant. A small star evolves into a white dwarf; a big one bids farewell in a supernova explosion and collapses into a neutron star or a black hole. (NASA and Night Sky Network)

Big stars can cook ever more complex elements because they have the mass to keep applying the needed pressure and heat. But only up to a point. With the creation of iron, the game's over. Iron is stable and doesn't "burn" like the others, so fusion shuts off in the core.

Up to this point, the star has survived by holding gravity at bay. Gravity wants to draw the star together, to crush it. Energy from fusion pushes back. Once that energy source is exhausted, there's no more pushback, and gravity gains the upper hand. The star implodes, collapsing in on itself. When all that material reaches the core, it "bounces" off that newly forged iron heart at close to light speed, creating a rebounding shock wave that blows the star apart in a supernova explosion.

Good news! That won't happen to the sun. While massive by human standards, it's almost insignificant compared to a supergiant star. The sun is currently burning hydrogen into helium, which slowly accumulates in the core. Gravity will keep putting on the squeeze, compressing the core until the helium reaches about 100 million degrees Fahrenheit (55.6 million degrees Celsius), hot enough to fuse it into carbon and oxygen. That's expected to happen about 5 billion years from now. If we could travel to that distant day and peer inside the sun, we'd see a

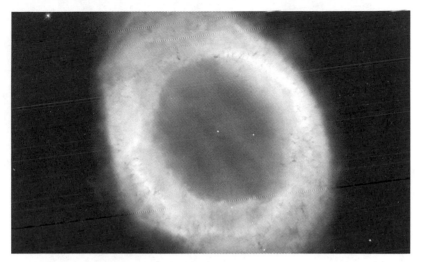

On its way to becoming a white dwarf 5 billion years hence, the sun will shed its outer envelope as a spherical cloud of glowing gas called a planetary nebula that may resemble the Ring Nebula pictured above. (Hubble Heritage Team AURA-STScI-NASA)

carbon–oxygen core surrounded by a shell of burning helium surrounded by an outer shell of burning hydrogen—Russian matryoshka dolls on a stellar scale.

The extra heat generated by helium burning will cause the sun's outer layers to expand outward, cool and redden, turning our yellow star into a red giant. How big it will get is still a matter of conjecture, but long before it reaches maximum girth, the heat radiated from its outer layers will roast the Earth, turning it into a blisteringly hot desert. As the sun expands further, there's a fair chance it will engulf and incinerate our planet. And that will be that.

Or maybe not. My guess is that whatever remains of our species, assuming we're smart and lucky enough to survive that long, will have departed long before the expected catastrophe. Seeing the inevitable, our descendants would gather what they could of Earth's life and shove off to more clement places. Perhaps even Mars.

As we left our fated planet, the sun would then transition from red giant to white dwarf. Like a creature in metamorphosis, it will shed its outer layers in rings and whorls of glowing gas called a *planetary nebula*, exposing its core, now transformed into a superhot, superdense white dwarf. Planetary nebulae are among the loveliest sights in the sky. When I look at them though my telescope, I think about the planets and living things that have or will pay the ultimate price.

187

SOLAR ECLIPSES PRODUCE DANGEROUS RAYS THAT WILL BLIND YOU

+ ✦ +

Except for a few precious minutes during totality, the sun is *always* dangerous to look at—in or out of eclipse. We sometimes forget, but that big, shiny ball up there is a giant nuclear fusion reactor 864,000 miles (1.4 million km) across. We'd all be toast were it not for the 93 million miles (150 million km) that stand between us and its fury.

Still, sunlight can hurt us if we're not careful. The sun radiates energy across the entire electromagnetic spectrum, from radio waves to high-energy X-rays. Our atmosphere is completely transparent to visible light, a good chunk of the radio spectrum and slivers of infrared and ultraviolet (UV), the light that makes black-light posters glow.

Solar UV light is also responsible for aging our skin and causing painful sunburns. Staying outside in intense sunlight for long periods of time can cause photokeratitis, a sunburn of the cornea (the outer, transparent layer of the eye) and the membranes lining the eyelids. Steady exposure to sunlight over the years increases our chances for cataracts and other eye ailments. Simple precautions like wearing sunglasses and a hat can help prevent these troubles.

When you feel the heat of the sun on your face, you're sensing infrared light. Like UV, infrared (IR) is invisible. It lies just beyond the red end of the rainbow spectrum in the same way that UV begins just past the violet end. Infrared light was discovered in 1800 when English astronomer William Herschel used a prism to spread a beam of sunlight into a spectrum, and placed a thermometer under each color to measure

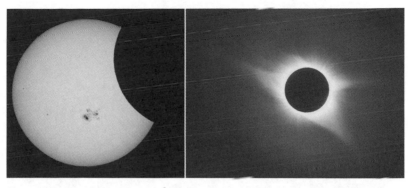

The partial solar eclipse of Oct. 23, 2014 featured a gigantic sunspot group. Right: When the moon completely covers the sun as it did during the Aug. 2017 total solar eclipse, you can safely look up and admire the sun's corona without a filter. (Bob King [left], Richard Klawitter)

its temperature. He noticed that it increased from blue to red, with the highest temperature recorded just beyond red, where *no light* was visible.

Visible, UV and infrared light account for 99 percent of the light that reaches the Earth's surface. Energy-wise, visible light contributes 42 to 43 percent, infrared between 52 and 55 percent and ultraviolet 3 to 5 percent.

The good news is that during solar eclipses sunlight is no different than it is out of eclipse. No special rays are produced, just the normal visible, IR and UV. What *is* different is our fascination with the sun during an eclipse and the temptation to take a peek without proper eye protection. On any other day, no one would deliberately stare at the sun, but during an eclipse, a few folks ignore their natural inhibitions and look anyway. A split-second glimpse won't kill you, but if you look directly at the sun even for a few seconds, your retinas can suffer permanent damage primarily from infrared light.

That's what happened to 26-year-old Nia Payne of Staten Island during the August 21, 2017, total eclipse. When the moon had covered about 70 percent of the sun, she stared at the thick solar crescent for about 6 seconds before deciding to cover her eyes. Still wanting to watch the eclipse, she borrowed a pair of what she thought were safe eclipse glasses from someone nearby and looked for another 15 to 20 seconds. Unfortunately, the glasses weren't the right type. At the time, she felt nothing because there are no pain receptors in the retina.

A park service employee at Arches National Park safely watches the May 2012 solar eclipse with approved eclipse glasses. (Neal Herbert)

But for the next two days she couldn't shake the sight of a dark "crescent" hovering in the center of her field of vision. When doctors performed detailed scans of her retinas, they discovered that the dark crescent was a mirror image of the one she saw during the eclipse, burned into her retina by the sun's infrared light. Six months after the eclipse, her vision compromised, Payne still struggled with reading and watching TV.

Outside of totality, the only time you can safely view a solar eclipse, take care to protect your eyes.

While sunglasses and UV coatings will reduce daily wear and tear on the eyes, they're useless for direct solar observation.

Always use an aluminized plastic Mylar or glass filter specially made for viewing the sun or a #14 welder's glass. These transmit just 0.0003 percent of the sun's light and provide safe and comfortable views. You can also project an image of the sun onto the sidewalk, a sheet of white paper or a white pillowcase using a homemade pinhole projector. Make a very small, neat hole in a piece of cardboard or paper plate, lower it a few inches above the ground and look beneath it to see a host of tiny suns projected below. Or use a kitchen colander with its hundreds of perfect holes—one of the best solar eclipse viewing aids around.

BETELGEUSE WILL GO SUPERNOVA AND RENDER EARTH UNINHABITABLE

+ + +

We like to worry about things. If all the people in the world were just la-di-da we would have been eaten by lions or overcome by natural disaster long ago. I guess the downside of knowledge is that the more we learn about the universe, the more things there are to worry about.

Take supernovae. If the possibility that one might explode near the sun and sear our defenseless planet with gamma rays is keeping you up at night, rest easy. For that to happen, a massive giant or supergiant star would have to lie within about 50 to 100 light-years of the sun. None do.

Betelgeuse is often cited as the prime example of a star likely to explode as a supernova in the near future. And what a sight it will be. By some estimates the star could become nearly as bright as the full moon but, unlike the moon, would appear as an intensely bright point source. Betelgeuse would cast sharp-edged shadows at night and shine brightly even when the sun is up. Let's hope you and I both are around when it happens.

Betelgeuse lies 642 light-years from Earth, so we're safe from its future fury, but there are closer supernovae candidates. Antares in Scorpius the scorpion is another red supergiant star located 550 light-years away. Both Antares and Betelgeuse are about 700 times the size of the sun, near the end of their fuel supply and could go supernova any time in the next million years. That could mean tomorrow night or in the year 1,002,019 C.E., give or take.

Other candidates include Virgo's brightest star Spica and the variable star IK Pegasi in Pegasus the flying horse. Spica is actually two stars in close orbit. Both are hot and massive, with the brighter one nearing the end of its life. More than

Betelgeuse is a bright supergiant star in Orion, the hunter, located to the upper left of the familiar 3-stars-in-a-row "Belt" asterism. (Bob King)

ten times as massive as the sun, Spica can fuse heavier elements right up to iron and implode as a supernova. At 250 light-years away, Spica is considerably closer than the two red supergiant candidates but still well beyond the range of threatening Earth. No need to run for the tin foil hats just yet.

If you recall from the earlier section (page 183), supernovae come in two distinct flavors: massive star explosions and white dwarfs that over-snack on their stellar companions. IK Pegasi, located just 150 light-years away, is a good example of the latter and the closest star that could potentially erupt as a supernova. Still, you needn't hold your breath. This one will develop over a much longer timeline with a surprising and safe ending.

Like Spica, IK Pegasi is a stellar duo, with two suns in close orbit about the other: IK Pegasi A, a normal white star that burns hydrogen like the sun, and IK Pegasi B, a massive white dwarf. For now, each is separate, connected only by their mutual gravitational attraction. But as A evolves into a red giant, its outer envelope will expand and approach the dwarf, which will commence siphoning off some of the star's hydrogen and helium gases. Once enough material has accumulated on the dwarf's surface, a nova may ensue. Or it could grab overmuch gas to exceed the Chandrasekhar Limit and explode as a supernova.

Gamma-ray bursts (GRBs) are blasts of energetic gamma-rays lasting from less than a second to several minutes and are among the most powerful events in the universe. If a jet were aimed at Earth, we'd see a brief but powerful gamma-ray flash. (ESO/A. Roquette, CC BY-SA)

Because millions of years must still pass before the star becomes a red giant, there's a good chance that IK Pegasi will have moved several hundred-light years farther away from our solar system and pose a miniscule threat.

We appear to be safe from ordinary supernovae, but possibly not from a gamma ray burster or GRB. A GRB forms when an unusually massive star explodes as a supernova, but instead of collapsing and rebounding in a star-shredding cataclysm, the outer layer explodes and a black hole forms at the center. Energy released in the explosion gets shot out as twin beams of gamma rays, the most deadly form of radiation known, that can skewer a hapless planet in its path.

WR 104, a massive binary star located about 8,000 light years from Earth, has GRB potential. When the brighter of its two suns goes supernova, there's a small possibility it could beam gamma rays directly at Earth, destroying about 30 percent of the ozone layer and doing other nasty chemical damage to the atmosphere. Not the end of the world but certainly unpleasant.

Or the beam could simply swing by and miss us. Who knows and frankly, hardly worth the worry. Sleep well.

YOU CAN HEAR SOUNDS IN OUTER SPACE

✦ ✦ ✦

In outer space there is no air, so there is no sound. Sound requires a medium like air to get from its source to your ears. When you clap your hands, it causes the air to vibrate. Air carries those vibrations into your inner ear, where they vibrate your eardrum, which the brain interprets as sound. Because space is a near-perfect vacuum, it can't transmit sound.

This is very different from light and its many forms like radio waves, X-rays and ultraviolet rays. Light waves are electromagnetic waves—not pressure waves like sound—and don't require a medium for transmission. Mission specialists stay in touch with spacecraft in the far reaches of the solar system using radio waves, a form of light. Sensitive cameras trap the light of galaxies that crosses billions of light-years.

Spaceships get blown up all the time in sci-fi movies and TV shows. Much of the impact of these scenes is visual, but sound plays a key role. Imagine a ship exploding in outer space in complete silence—what fun would that be? That's why producers add traditional sounds to battle scenes and destroyed planets. But a real explosion in space would be utterly silent. And forget the fire and fury, too. There might be an extremely brief flash as oxygen in the ship's airlock ignited, but once that was used up, we'd see only jagged shards of metal and glass flying outward from a silent blast.

If we were on a neighboring ship, we might get to hear the sound of shrapnel from the explosion crashing into the hull because metal and other solid materials transmit sound. As a kid, I used to press my ear on one of the rails of a railroad track to listen for the first rumble of a distant train. Then we'd place pennies on the rails and hide until the train passed. If we were lucky, our pennies would be rolled thin by the weight of the passing cars.

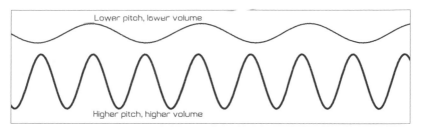

Sound waves are vibrations that need a medium to be heard whether that be air, water or solid metal. Widely-spaced sound waves (top) produce a deep bass sound. Closely-spaced waves have a higher pitch. (Pluke / CC 0)

An imaginary spaceship exploding in outer space creates no sound, and any flames produced are small and quickly dissipate from a lack of oxygen. (Gary Meader)

Even a supernova explosion would produce nothing more than a deafening silence. If you pulled up to a reasonably safe distance from the star and watched the show, you'd instinctively recoil at the sight, but even though lots of gas would be expelled in the eruption, it would rapidly thin out in the vacuum of space and hardly register a pop to human ears. Our ears require a medium and one that's dense enough to carry sound. A few atoms here and there won't cut it.

Liquids and solids also vibrate and transmit sound, as anyone who's ever ducked their head underwater knows. When I'm snorkeling, I enjoy listening to the sounds of sifting sand and rising air bubbles. The speed of sound varies depending on the medium. Sound travels slowest in air at 767 mph (1,234 kph), faster in water at

3,320 mph (5,343 kph) and fastest in solids, racing through iron at 11,450 mph (18,427 kph) or 15 times air speed. That may sound strange until you remember that the atoms and molecules in solids are packed much more tightly than they are in air. Vibrations literally go right through them. With gases, the molecules are much farther apart, so it takes more time for vibrations to go from one place to the next.

It's been said that in space no one can hear you scream. And while true, most sensible astronauts would put on a space suit first and communicate (or scream if needed) using radio. Within the microenvironment of your suit, you can hear your own movements, the sound of your breath and the mechanical noises from the cooling system and oxygen supply. If a fellow astronaut knocked on your helmet you'd hear that too, because vibrations would be transmitted through your mask to the air inside the helmet and into your ears.

Nebulas are made of gas. Would it be possible to hear sounds or someone talking if you could remove your helmet and listen closely? Let's check. Interstellar space contains about one atom per cubic centimeter (cm^3). That's a cube 0.40 inch on a side, the size of a small gaming die with just one atom tucked inside. The best vacuum we can create on Earth contains around 100 particles/cm^3. That's pretty good but still 100 times more cluttered that a cube of outer space. In gas clouds like the Lagoon and Eta Carina Nebulas, typical densities are around 100 to 10,000 particles/cm^3. The density of Earth's atmosphere at sea level, where most of us do our shouting, is 27,000,000,000,000,000,000 quintillion molecules/cm^3, some 15 orders of magnitude greater.

Very little research has been done on the minimum density of molecules necessary for our ears to detect sound, but there's no question sound would propagate across a nebula. The only rub is that the waves would be so weak our ears wouldn't be sensitive enough to detect them. Perhaps in the nebula's densest region, with a massive woofer pumping out Queen's *Another One Bites the Dust* bass line, we might detect the throb with an ultrasensitive recording device. Low-frequency (bass) sounds have longer wavelengths than tinkly bells and are better able to bridge long distances between scant concentrations of atoms.

If I might share a piece of sound advice, the next time you're watching a battle scene set in space, mute the volume for a more realistic experience.

MATTER IS EITHER A SOLID, LIQUID OR GAS

It's bad enough that normal matter, the stuff of cars and couches, only accounts for 5 percent of all the matter in the universe. The rest is dark matter that binds together galaxies and galaxy clusters. Its nature remains unknown. But of that 5 percent, 99.999 percent isn't even the familiar stuff of everyday life.

We learned as kids in science class that matter comes in three states—solid, liquid and gas. In a solid, atoms or molecules are close together and fixed in place like planks in a scaffold. In liquids, they're free to slide around and take the shape of the container holding them. Gases are looser yet, with molecules neither close together nor fixed in place. Gases expand to fill the entire volume of whatever container encloses them.

You can convert one state or phase to the other by adding or removing heat. Water is a good example. If you chill liquid water to its freezing point, H_2O molecules will join together to form microscopic hexagons, which bond with each other to form a hexagonal (six-sided) crystal lattice. This is the underlying reason snowflakes have six sides or points. If you boil liquid water to make tea, you'll also make steam, the gas phase of water.

In spite of their familiarity, solid, liquid and gas are little more than a drop in the bucket when we consider the universe as a whole. If you want to see the most common form of matter, just look up at the sky on a clear night. Every star and many nebulae are composed of matter's fourth state, called *plasma*. Closer to home, you'll also find plasma aglow in fluorescent bulbs, the aurora and welding arcs. Given the estimated two trillion galaxies in the universe, each of which glitters with hundreds of thousands to trillions of suns, plasma by far makes up the lion's share of visible matter.

Water as a solid is most familiar as ice cubes but it also creates the crystal filigree of a snowflake. (Bob King)

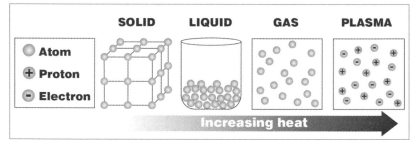

On Earth we're most familiar with solid, liquid and gas forms of matter. Yet the most common is the 4th state or plasma, the stuff of neon signs and stars. (Gary Meader)

Plasma resembles a gas in that it has no shape and can expand to fill the available space, but it differs because the atoms in plasma have an electric charge. At the atomic level, familiar matter is electrically neutral—each atom or molecule's electrons stay bound to the nucleus. Plasma forms when a gas has been heated to such an extreme temperature that the electrons leave their atoms and wander freely. Without their negatively charged wrappings, the atoms become bare, positively charged nuclei called *ions* swimming in a sea of free electrons. It's like a nice, steamy bowl of meatball soup where the meatballs (ions) bob about in the hot liquid (electrons).

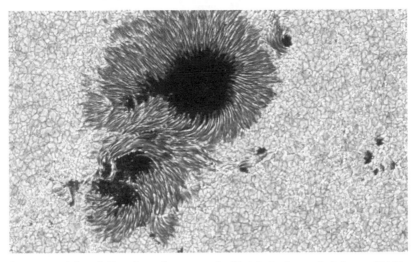

Sunspots and the cell-like solar surface are made of hot, ionized gas called plasma. (NASA / JAXA)

What makes plasma unique is that charged particles conduct electricity and react to and produce electric and magnetic fields. When plasma from the sun—in the form of positively charged protons (nuclei of hydrogen atoms) and electrons— blows by Earth it can hook into and be guided by Earth's magnetic field straight into the upper atmosphere. That would never happen with an ordinary, neutral gas. When the solar ions slam into oxygen and nitrogen atoms, they dislodge their electrons, creating a momentary plasma. When electrons rejoin their parent atoms, energy is released as the pink and green light of the aurora.

Solar plasma is made mostly of ionized hydrogen (a fancy name for a single proton stripped of its single electron) and helium along with free electrons. Hot plasmas laced with their own magnetic fields bubble up to the surface from deep within the sun. As the sun turns on its axis, those fields are spun ever tighter until they're concentrated enough to produce sunspots and their associated magnetic storms called *flares*. Flares and other solar outbursts send clouds of plasma into space and spark the aforementioned northern and southern lights.

Plasma sounds a little like electricity, but the two are different. Electricity is the flow of negatively charged electrons from one atom to the next, like people (atoms) passing along buckets (electrons) in an old-fashioned bucket brigade.

Plasma consists of a nearly equal number of negatively charged electrons and positively charged ions flowing together side by side, making it *overall* neutral yet still able to conduct electricity.

High voltages can also make plasma. That's how neon signs work. An electric current passes through the gas, separating electrons from atoms; when they recombine, ultraviolet light is released. The light strikes a luminescent chemical called a *phosphor* that coats the inside of the tube. Depending on the type of phosphor used, a wide variety of colors result. In a plasma TV, an electrical current passes through millions of microscopic cells filled with either xenon or neon and are individually coated with phosphors that glow red, blue or green. When an electric current zaps the xenon, the gas becomes an electrical-conducting plasma and emits UV light, exciting the phosphors that together create an image.

There are even more states of matter, all of them exotic or hypothetical. Examples include Bose-Einstein condensates (a gas of subatomic particles cooled to near absolute zero); neutron degenerate matter (superdense gas composed of neutrons inside a neutron star) and quark-gluon plasma (produced in the extreme heat a tiny fraction of a second after the big bang).

If plasma has your head spinning, stop by your local brewpub and pull up a chair under the glow of the neon sign hanging in the window and you'll see the light.

CONSTELLATIONS ALWAYS KEEP THEIR SHAPES

+ ✦ +

Night after night, season after season and year after year, we can count on the constellations to keep their shapes and stay put. Planets, on the other hand, hurry around the zodiac like runners on a racetrack. One year, Jupiter gleams from Aquarius. Four years later, it's three constellations over, in Taurus. Planets move because they revolve around the sun and they're close to us. Stars are also in motion, but they're so incredibly far away that they hardly stray from their appointed spots. If you shrunk the sun down to the size of a basketball, the Earth would be a BB just 31 feet (28 m) away. At the same scale the next closest star, Alpha Centauri, would be a pair of basketballs (it's a close double) 4,300 miles (6,920 km) away!

Cassiopeia will spell the letter W for centuries to come. Generations of our descendants will marvel at the simple symmetry of Orion's Belt. Constellation patterns seem fixed for eternity, but are they?

Arcturus, the brightest star in the constellation Boötes, twinkles high in the southern sky at nightfall in June. Every night, Earth's rotation "drives" the star from east to west across the sky. Of course, we know that Arcturus itself isn't moving—the Earth is. But like every star, Arcturus has its own intrinsic or *proper motion*. As it orbits about the center of the Milky Way Galaxy at 76 miles per second (122 km/sec), it inches 2.3 arcseconds to the southwest every year. That's equal to 1/800th the apparent diameter of the full moon. Talk about a slow crawl. Even if you lived to be 100, you'd never notice a change in its position with the naked eye.

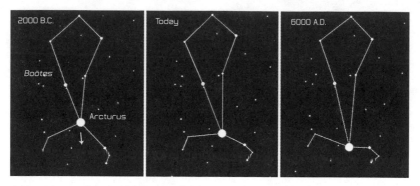

Arcturus is one of the few bright stars that has moved noticeably since the days of ancient Greece and Rome. As it continues to travel southwest, it will slowly distort the outline of the constellation Boötes. (Bob King)

But those arcseconds add up. If you were to make a careful drawing or take a photograph of the star when you were 20 years old and then redraw its position 40 years later, Arcturus would have moved 92 arcseconds, or about 1/20 of the moon's diameter, a distance obvious at low magnification in a telescope.

If you could jump in a time machine and zoom back to ancient Rome, would you notice that Arcturus had moved? Yes! Assuming you knew your constellations, you'd clearly see that the star had shifted a full degree (equal to two side-by-side full moons) northeast of where it is today. Sirius would also be slightly out of place, about one full moon northeast of its current location. Otherwise, past and present skies would appear identical, at least to the casual observer.

Edmund Halley of Halley's Comet fame discovered proper motion when he noticed that the positions measured for Sirius and Arcturus in ancient Greek star catalogs were different from those of his time. Both stars are relatively close to the Earth—8.6 and 37 light-years, respectively—so their apparent motion across the sky reveals itself over a shorter amount of time than it does for most stars. Arcturus also happens to be moving almost perpendicular to the sun, making its travels even more obvious over a short period of time compared to a star moving in tandem with the sun. Since Halley's day, astronomers have discovered dozens of even faster-moving stars, but they're only visible in a telescope. Some scoot along quickly enough for us to see movement in only a year's time.

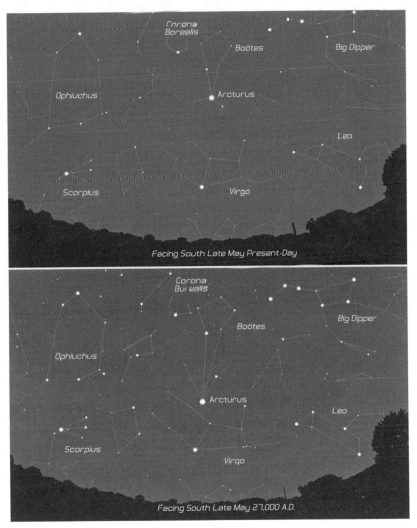

We'd all be scratching our heads if transported 25,000 years into the future. Because of stellar motion many constellations would be unrecognizable. (Stellarium)

Using a planetarium-style software app, I checked and discovered that over the past 5,000 years—the span of written history—the bright stars Sirius, Arcturus, Altair, Procyon and Alpha Centauri have moved enough that if an alert skywatcher were transported to 3500 B.C.E., she'd easily notice the shifts. A modest number of fainter, less obvious stars along with a host of telescopic suns would also have changed position, but the constellations would still appear much like we see them today.

Let's hit the accelerator and fast forward 25,000 years into the future. Looking up, we'd be shocked. Many of the constellations' shapes would appear distorted and stretched as if seen through a fishbowl. Thankfully, a few old standbys would help us get our bearings: Orion, the Big Dipper and Taurus would still look familiar, if distorted. And 100,000 years from now, with few exceptions, the sky will appear completely alien. Because so many stars will have migrated to new spots, we'd have to start from square one and rebuild the sky with new constellations. Orion will be one of the few that will still keep its shape on that distant date. A little stretched, maybe, but even the belt will remain recognizable. I suspect the reason is because the sun and many of Orion's brighter stars are traveling in tandem with the sun around the galactic hub. Relative to each other, they're moving slowly.

Across the sweep of time, all constellation figures will fall away. Bright stars will fade as they recede from the sun; distant ones will approach and grow brighter. Speed up the arrow of time and the steady stars stream like meteors through the darkness.

I've always loved time travel ever since seeing the movie *The Time Machine*, based on the book by H.G. Wells. It took years, but I finally have my own personal time machine. You do too. Download a free copy of Stellarium (see Resources below), set the date to the past or future and go for a ride!

Resources
* Stellarium: http://www.stellarium.org

BLACK HOLES SUCK

+ ✦ +

People like to say that black holes suck, but that's the *last* thing they do. If you want something that sucks, look no further than your vacuum cleaner. It works by suction. A powerful motor sucks in air along with dirt and dust and collects it all in a bag. Instead of *pulling* things in, objects *fall* into a black hole if they stray too close. Black holes aren't nearly as willful as that noisy machine in your closet.

A black hole forms when an extremely massive star—more than about 20 times the mass of the sun—runs out of nuclear fuel. Stars spend their lives arm-wrestling with gravity by fusing atoms in their cores to make energy. The outward heat and pressure from nuclear "burning" counteract gravity's crushing grasp. In a small star like the sun, once its nuclear fuel is used up, gravity will compress the core into a small, extremely dense little star called a white dwarf. Electrons whirling around the atoms in the dwarf are squeezed so close together they create a pressure that resists further crushing.

But in massive stars, once the fuel gauge hits empty and burning ends, the star implodes and the core rapidly contracts. If the core's mass is about three times the mass of the sun or greater, it can't resist further collapse and will keep on shrinking and shrinking until it becomes a mathematical point of infinite density called a *singularity*, surrounded by a boundary called the event horizon. A typical core collapse results in a black hole with an event horizon about 11 miles (18 km) across. These are called *stellar mass black holes*. At the same time the black hole is forming, a shock wave rebounds outward from the core and tears the star apart in a supernova explosion.

When you compress that much matter into a tiny space, even light is held back by gravity and can't escape. The event horizon marks the black hole's "edge." Here, an object would have to travel at the speed of light to escape falling in. Since we know that's not possible, if you nudged your spaceship up to this line, there would be no escape. Inside the horizon, both matter and light are captive.

Not all supernovae produce black holes, only those involving the most massive stars. A black hole has no surface. Instead, it's a region of space bounded by its

This artist's impression shows a supermassive black hole surrounded by a disk of material that's heated to glow as it falls into the black hole. Beams of particles and radiation focused by the disk's powerful magnetic field blast away from the object at near-light speeds. (ESO / M. Kornmesser)

event horizon. You can safely orbit a black hole if you remain beyond the event horizon, but cross it and you'll ultimately become one with the singularity. According to NASA's Hubble Telescope site, Earth would have to stray within about 10 miles (16 km) of a black hole to be in danger of falling in.

Black holes aren't picky about what they eat. The usual diet is what floats throughout all of space—hydrogen and other gases, cosmic dust and the occasional stray star. No matter what falls in, all the material gets funneled into the singularity. While the singularity remains a mathematical point, the more matter the hole gobbles, the larger the event horizon and the more massive the black hole becomes. For every solar mass a hole swallows, its event horizon grows by 1.9 miles (3 km).

That brings us to the second type of black hole, the supermassive ones that lurk in the centers of many galaxies, including the Milky Way. While stellar mass holes contain from 5 to maybe 30 suns of material, the monster ones can pack away millions to billions' worth of suns with event horizons as wide as the solar system. Astronomers are still trying to figure out the origin of supermassive black holes. Collapsing star clusters or the merger of many smaller holes into one large one are the current best explanations.

For years, astronomers have only been able to infer the existence of black holes by their effects on other objects orbiting near them. But no more. In April 2017,

The Event Horizon Telescope (EHT) captured this first-ever image of the supermassive black hole in the galaxy M87 in Virgo in early 2019. Gases falling into the hole are heated to high temperatures and create the glowing wreath. The hole's event horizon measures about 25 billion miles (40 billion km) across. (Event Horizon Telescope Collaboration)

researchers ganged together eight radio telescopes spread across the globe into one ginormous, Earth-size instrument called the Event Horizon Telescope (EHT). Combined, the eight-eyed animal could discern details 2,000 times finer than the Hubble Space Telescope. They focused it on the center of the galaxy M87, located 54 million light-years away in the constellation Virgo. Two years and five petabytes (one petabyte equals a million gigabytes) of data later, the EHT team released the first-ever image of the supermassive black hole in the galaxy's center.

Relativity theory predicted we should see a round, black emptiness silhouetted against hot, glowing gases swirling past the event horizon. Incredibly, that's exactly what the image revealed! It showed the black, round shadow of the event horizon framed by hot, glowing gases swirling down into the hellish blackness.

Gravity—a black hole's iconic attribute—gives them away. Earlier, astronomers tracked down M87's supermassive black hole by studying the motions of clouds of gas in the center of the galaxy. The same way Earth is tethered to the sun by gravity, something also had to keep those clouds in orbit. That something proved to be 6.5 billion times more massive than the sun and concentrated in a space some 24 billion miles (38 billion km) across, or about a quarter the size of our solar system. *And* it was completely invisible, as if nothing were there. Man, this thing almost screamed supermassive black hole.

Astronomers have observed similar motions in the center of the Milky Way, with stars whipping around an invisible "nothing" at incredible speeds. Our galaxy's supermassive black hole is a baby compared to M87's, with a mass of about 4.3 million suns and a diameter of 93 million miles (150 million km), about the same as the distance between the Earth and the sun. Only black holes are able to wield such force and yet remain unseen. Despite its smaller physical size, the Milky Way's black heart is more than 2,000 times closer, so its *apparent* size is similar to M87's monster. During that same week of good weather in April 2017, the EHT team also gathered a slew of data on Sagittarius A* (A* is pronounced "A Star"), the site of our galaxy's supermassive black hole. As I write, they're hard at work on that image and may already be published.

When gas and dust fall into a black hole, they enter with a little bit of sideways motion and get swept into a disk or whirlpool orbiting the hole. Closer-in matter moves faster, grinding against material further out, which orbits more slowly. Friction heats them to millions of degrees. If you heat matter to high temperatures, it radiates not only heat but also light, including extremely energized light like the X-rays a dentist uses on your teeth during a check-up.

Astronomers have detected a number of *X-ray binary stars*, where a black hole and a normal star tightly orbit each other. Like the nova scenario described earlier, where a white dwarf funnels material from its companion, the black hole siphons gas which radiates X-rays on its way to the event horizon. Astronomers use orbiting X-ray telescopes to study these odd couples. By measuring the speed of the orbiting star, we can determine the mass of the object pulling on it. When it exceeds several suns and hides from sight, we know we've nabbed another black hole! Similar incandescent gases revealed the presence of M87's black hole.

You could turn the sun into a black hole if you crushed it into a sphere 3.7 miles (6 km) across. Its gravity would then be so powerful that not even light could escape. Just like that, the sun would wink out. Earth would continue orbiting the now "dark sun" exactly as it did before. To make the moon a black hole, you'd need to pack all its matter into a sphere the size of a poppy seed. Although completely invisible, it would continue to orbit the Earth and raise tides as usual. Sorry, though—no more eclipses!

No force we know of could do this to the Earth, sun or moon. Only truly massive cosmic objects hold the key to the gravitational "dark side." And not to worry about Earth falling into a black hole anytime soon. The nearest one, A0620-00, is located in the V616 Monocerotis system about 3,300 light-years away. There, an orange star loops around an invisible companion with a mass of 6.6 suns. We're safe!

AS THE UNIVERSE EXPANDS, GALAXIES SPEED AWAY FROM EACH OTHER

+ ✦ +

Space used to be so simple. Just a whole lot of static nothingness. It's not like that anymore. Einstein, with help from his former math teacher Hermann Minkowski, combined the three dimensions of space with time into a single, four-dimensional framework called *space-time*.

We also used to think time ticked at the same rate for every observer anywhere in the universe. But time, like space, is flexible. As Einstein showed, the amount of *space*, or distance, between two objects and the *time* it takes to get there vary according to your speed, especially at velocities approaching that of light.

Alpha Centauri is 4.4 light-years or about 26 trillion miles (42 trillion km) away. Even light takes 4.4 years to get there. But if I jump into an ultrafast spaceship and throttle up to 90 percent light speed, I'd arrive in just 2.12 years. Meanwhile, a person tracking my progress back on Earth would report that it took me 4.86 years *by their clock* to make the journey. Did I just travel faster than light?

No! As we learned earlier, matter can't travel at light speed. What happened instead was that space *shrunk*, reducing the distance to the star. A double-check of the odometer confirms that my trip came to 1.9 light-years, or less than half the distance had I traveled at a more conventional speed.

Time's flexible rate highlights a disturbing consequence of near light-speed travel: 2.12 years of time passed for me but 4.86 years ticked by for my buddy back on

The gravity of the sun (left) and the Earth warp the fabric of space and time, or space-time, shown here as the curved grid. The dimples are similar to those in a trampoline when stood upon. (Courtesy T. Pyle/Caltech/MIT/LIGO Laboratory)

Earth. If I turned around and zoomed back to Earth, he'd be 5½ years older when we met again! If I increased the ship's velocity to a tiny fraction shy of light speed, so that I'd arrive at Alpha Centauri in 10 minutes flat, 1 million years would have passed for my (now) deceased friend.

Time dilates (slows down) and distance contracts from the perspective of someone traveling at a significant fraction of the speed of light, making travel times and distances vary depending on your frame of reference—in this case, between Earth and a fast spaceship. This brief foray into space-time prepares us to better understanding speeding galaxies.

Space and time came into existence about 13.8 billion years ago at the beginning of the universe in the big bang. The temperature inside the initially microscopic universe measured in the trillions of degrees. But by the end of the first 3 minutes, it had expanded and cooled enough for hydrogen and helium to form, the dominant elements of the universe to this day. Clouds of these elements coalesced to build the first generation of stars roughly 250 million years after the big bang. Not long after, stars gathered together into galaxies.

In the early twentieth century, astronomer Vesto Slipher of Lowell Observatory in Arizona had time to focus on his passion—measuring the speeds of galaxies. His

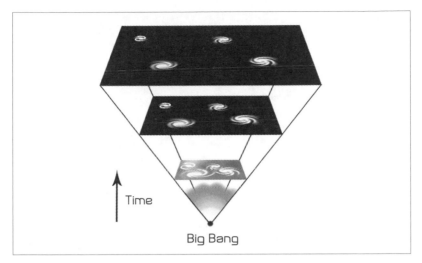

Time

Big Bang

The space between the galaxies has been expanding since the origin of the universe in the Big Bang. This creates the appearance of galaxies racing away from one another when they're really just going along for the ride. (CC BY-SA 3.0 / Wikipedia)

data revealed that the Andromeda Galaxy (then called a nebula) was approaching us, while a number of other galaxies were moving away. In the 1920s, Edwin Hubble combined Slipher's data with his own observations, finding that nearly every galaxy beyond the nearby "Local Group" was receding from the Milky Way. And the greater the galaxy's distance, the faster it hurried away. The year was 1929, the start of the Great Depression and the same time that Hubble came to the astonishing conclusion that the *universe was expanding*.

Expansion only happens across huge distances. You and I aren't expanding (at least not cosmically!), nor is the Earth, solar system or Milky Way Galaxy. Locally, gravity holds things together, like the planets to the sun, the sun to the galaxy and nearby galaxies to one another.

Not until we look beyond galaxy clusters does universal expansion play its hand. Astronomers determine the expansion rate by measuring the speeds at which galaxies appear to be receding from us. To our best knowledge, it's 43 miles per second (70 kps) for every 3,260,000 light-years we peer deeper into space. So, a galaxy 3.26 million light-years from Earth would be moving away at 43 miles (70 km) a second. The speed of a galaxy at twice that distance would double to 86 miles (140 km) a second.

I keep talking as if galaxies were flying off into deep space like Frisbees gone wild, but in fact, they're *not moving at all*. Instead, space between the galaxies is stretching like two people pulling on either end of a rubber sheet with the galaxies just going along for the ride. While rubber can be handy for understanding space, a loaf of uncooked raisin bread works even better.

Lets pretend the loaf represents all of space-time with raisins for galaxies. As the dough bakes and expands, the raisins move away from each other. If each raisin *doubles* its distance from the others during the bake, then a selected raisin—we'll call it Ed—positioned ½ inch (1.2 cm) from its neighbor at the start, will be separated by a full 1 inch (2.5 cm) when the loaf is fully baked. Another raisin 3 inches (7.5 cm) from Ed will double its distance to 6 inches (15 cm). From Ed's perspective, the nearby raisin moved ½ inch (1.2 cm) during the bake, but the further one moved *3 inches* (7.5 cm) during the same interval, making it appear as if it moved much faster than Ed's neighbor.

Ed pauses to reflect: "Hmmm. I see that all the raisins are moving away from me, and the farther they are, the faster they move." This is exactly what we observe in the real universe, where space stretches rather than dough. Ed's friend Edna is stuck in a different part of the loaf and keeps an eye on a different set of raisins during the bake. She notices the exact same thing: all the other raisins are moving away from her, and the farther away they are, the faster they move. For any raisin anywhere in the loaf, the scene is identical. Galaxies and galaxy clusters are racing away from each other as the dough (space) expands between them. Because there's an excellent chance the universe is infinite, there is no center to this bread loaf. No edge. No privileged spot. All viewpoints are equal.

While matter can't travel at or faster than the speed of light, space has no such restriction. Using the expansion rate described earlier we find that galaxies that are 14.02 billion light-years from us are receding at faster than the speed of light. The visible universe constitutes what astronomers call the Hubble sphere, an imaginary sphere beyond which galaxies are receding at faster-than-light speed and for now remain invisible.

Before the universe began, there was only probability and possibility. But once the microscopic singularity that would evolve into the present-day universe appeared, space has been stretching ever since.

WHEN GALAXIES COLLIDE, STARS CRASH TOGETHER

A galaxy is an enormous collection of millions to billions of stars and their orbiting planets, along with gigantic clouds of dust and gas, star clusters, nebulae and black holes. Galaxies are the fundamental building blocks of the universe. When you look out into deep space, it's nothing but galaxies as far as telescopes can see. Many gather into massive clusters with several thousand members held together by their mutual gravity in consort with dark matter. Others are more isolated or members of smaller clusters like our own Milky Way, a member of the Local Group, a pack of 50-plus galaxies in a space about 10 million light-years across.

Space is big, but not so big that galaxies don't occasionally collide or pass so close to one another that their mutual pulls distort their shapes. Recent analyses of galaxy surveys made with the Hubble Space Telescope and computer simulations reveal that anywhere from 5 to 25 percent of galaxies merge together after colliding with each other. Large galaxies merge with other large galaxies about once every 9 billion years, while small galaxies meld with big ones much more often.

Galactic small fry, called dwarf galaxies, are the most common and are thought to have coalesced to build the larger galaxies like the Milky Way and Andromeda. Some of them are as small as 350 light-years across compared to our galaxy, which spans some 100,000 light-years. Mergers are most common in rich galaxy clusters where bigger galaxies "cannibalize" smaller ones, growing to immense sizes. The Coma Cluster, one of the richest, has at least 1,000 members. Located 322 million light-years from Earth, it takes its name from its home constellation, Coma Berenices.

The beautiful barred spiral galaxy NGC 1300, located in the constellation Eridanus the river, is about 110,000 light-years across, or about the same size as the Milky Way. (NASA, ESA, and The Hubble Heritage Team [STScl-AURA])

In this photo of the Arp 194 galaxy group, two galaxies (at left) are in the process of merging into one. Amid the confusion of spiral arms we can still make out their bright centers. A third much smaller galaxy hovers above the pair, while a fourth galaxy appears at right. Millions of newborn stars spawned by the collision create a stream of bright star clusters in its wake. (HST CC BY 2.0 ESA NASA processing by Judy Schmidt)

With galaxy mergers so common, you'd think there would be a lot of fireworks when stars in crashing galaxies slammed into one another. Fireworks *do* happen, but not because of stellar collisions. In the Milky Way, the average distance between stars is about 5 light-years, or 30 trillion miles (48 trillion km), a little more than the distance between the sun and Alpha Centauri. Picture each star centered in a starless sphere 5 light-years across. That's a lot of empty space. Stars come in many sizes, but even the largest ones with diameters around 2 billion miles (3.2 billion km) would occupy only the tiniest fraction of that enormous sphere. Essentially, stars are microscopic points separated by great volumes of practically nothing. My friend Michael once reflected on the human presence in the great immensity of space and summed it up nicely: "We're almost *not here* in comparison."

As one galaxy catapults through another, there's so much empty space between the stars that collisions are rare. For excitement to happen, you need something bigger than stars to slap together. Galaxies have just the ticket—gigantic molecular clouds composed of hydrogen (the most common element in the universe), dust, carbon monoxide and a variety of other molecules. Clouds hold together through self-gravity with diameters between 30 and 300 light-years and contain from 10,000 to 6 million suns' worth of material.

Denser regions within molecular clouds can collapse on their own under the force of gravity to birth new stars and star clusters, but colliding galaxies speed up the process, slamming and compressing the clouds, initiating wave upon wave of star formation. Each site looks like a pink firecracker, and colliding galaxies literally pop with them.

You may have heard in the news about an upcoming collision with big consequences for our own galaxy. Careful study of the motions of galaxies in the Local Group have revealed that the Andromeda Galaxy and the Milky Way, drawn by each other's gravity, are hurtling toward one another on a collision course at 250,000 mph (400,000 kph). At that speed, you could get to the moon in under an hour! They're still 2.5 million light-years apart, so it's going to be a while. Recent data from the European Gaia mission indicate that Andromeda will deliver a glancing blow to our galaxy 4.5 billion years from now. After a long and complicated gravitational "dance," the two galaxies will merge and trigger massive waves of new star formation. Should Earth still be around then—questionable because the expanding sun may have engulfed the planet by that time—what a spectacle our descendants will see. Throughout it all, there's an excellent chance no stars will be harmed in the making of the new mega-galaxy.

GALAXIES ARE SO FAR AWAY THEY MIGHT NOT EVEN BE THERE ANYMORE

Stars make the best time machines. The distances between the stars are so enormous that it takes years for their light to arrive. Traveling at 186,000 miles a second (300,000 km/sec), even sunlight takes 8.3 minutes to travel the 93 million miles (150 million km) to arrive at your window. We never see the sun in the here-and-now moment, only as it was 8.3 minutes ago. When Jupiter and Earth are closest, 33 minutes of light time separate the two planets.

If starlight takes years to reach our eyes, galaxy light takes *millions* of years. The Andromeda Galaxy, located 2.5 million light-years away, is the closest large galaxy to the Milky Way. Its light is truly ancient. Should you ever get an opportunity to see it, the light pinging your retinas that night departed the galaxy 2.5 million years ago. That's such a long time by human standards that some people wonder whether the galaxy—or faraway stars, for that matter—are still there or whether they've ceased to exist.

I'm happy to report the odds are excellent that every visible star *and* the galaxy yet live! Big numbers to us are small potatoes to long-lived entities like stars and galaxies. Andromeda has been around for about 10 billion years, and the Milky Way 13.5 billion years. Even after they collide and merge into one in 4.5 billion years, they'll continue to twinkle with stars old and new for tens of billions of years into the future.

The spectacularly luminous star AG Carinae shines from 20,000 light-years away in the Milky Way. It's classified as a Luminous Blue Variable. These rare, massive stars experience violent eruptions mimicking supernovae. (ESA/Hubble, NASA)

One definition of a "dead" galaxy is that it has run out of dust for new star formation. But even these, known as *elliptical galaxies*, are only "mostly dead" (to borrow a quote from the movie *The Princess Bride*). They lack spiral arms and the requisite dust and gas for new star formation, but they're still jam-packed with stars. A better definition of "dead" considers how long it would take for all the stars in a galaxy to burn out either in supernovae explosions or in more languorous fashion as slowly cooling red and white dwarfs. Then we're talking *trillions* of years.

The Andromeda Galaxy formed some 10 billion years ago and will be around for billions more, although in altered form once it merges with the Milky Way in the distant future. While there have undoubtedly been changes in Andromeda's appearance in the time it took light to reach our eyes tonight, 2.5 million years is just 0.00025 percent of the galaxy's current age. A drop in the bucket. If we could transport ourselves there right now, it would appear virtually identical to how it looks from Earth tonight.

Naked-eye stars are much closer than external galaxies. Are there any stars that may have changed significantly in the time it took their light to reach our eyes? Stellar lifetimes vary wildly. The most massive stars burn their fuel supply rapidly like a gas-guzzling luxury SUV and flame out as supernovae in just a few million years. At the other end of the food chain, red dwarfs burn their fuel so frugally that the smallest will shine for up to 10 *trillion* years, far longer than the current age of the universe.

So let's rephrase the question. Is there a supergiant star at least several million light-years away, far enough to be certain it's gone poof in a supernova explosion that's still visible today? Certainly none we can see with the naked eye. The farthest stars visible without optical aid are all closer than 10,000 light-years away—not enough look-back time to be certain they're gone. Even the most distant stars in the Milky Way are within 100,00 light years of the sun. Still too close! We'll need to look beyond the galaxy for our "living dead" star.

Luminous blue variables, or LBVs, are a possibility. These massive blue stars in other galaxies undergo violent eruptions, occasionally becoming bright enough to show as dim pinpricks of light in amateur telescopes. Because they ultimately explode as supernovae within a few million years, there are undoubtedly a few still shining that have long passed.

Another possibility is the star Eta Carinae (a naked-eye LBV visible in the southern hemisphere). It also may have imploded and left the scene, but because it's only 7,500 light-years away, chances are it's still there. Another somewhat better possibility is V4650 Sagittarii (also an LBV), which is 26,000 light-years away. Supernova candidates Betelgeuse and Antares also have potential. Each is expected to explode and may already have done so, but they're so nearby they may also still be shining.

As you can tell, it's nearly impossible to point to a current star and be certain it's no longer there. Stars have great longevity compared to the brief time allotted us on Earth. But as any star might tell you, it's not about how long you live—the important thing is to shine.

ACKNOWLEDGMENTS

+ ✦ +

I would like to thank my wife, Linda, who helped create my first office where I could write without taking over the kitchen, my former workplace. As well, I'm indebted to her practical suggestions for better living and the beneficiary of her lovely, handmade quilts.

Although my mother, Lorraine, passed on recently, I still feel her love and goodness and seek her internal guidance on how to be a good human. For fostering my inner skeptic, I must thank my father, Bill, who was critical of everything. I'm hoping the good part of that rubbed off on me, helping to sharpen my thinking and develop a scientific perspective.

I'm indebted to Gary Meader, graphic artist at the Duluth News Tribune, for his excellent and timely contributions to this book as well as all my former colleagues at the newspaper for their cheer, wit and ceaseless pursuit of the facts along with the wonderful stories and photos they share with the world.

I also want to thank my daughters Katherine and Maria for their love; my brothers Mike and Dan for their wicked humor and unwavering support; Sally King for her friendship and generous spirit; Ryan King for his musical gifts; Nova King for her passion; my friend Rick Klawitter for his photography and deep discussions; Phil Plait (a.k.a. The Bad Astronomer) for helping me understand tides; Valerie Blaine for her friendship and shared love of nature; Sam Cook for friendship and great writing; Roy Hager; Glenn Langhorst; the staff at *Sky & Telescope*; Fraser Cain, founder of Universe Today; the late Carl Sagan, a candle that still illuminates the darkness; and fellow author Nancy Atkinson for her friendship and support.

Finally, I'm indebted to my editor Marissa Giambelluca, Meg Baskis and all the fine people at Page Street Publishing for their help, support and expertise in making this book a reality.

ABOUT THE AUTHOR

Bob King loves the night sky and hasn't stopped looking up since he was ten years old. Born in Chicago, he grew up in nearby Morton Grove. When he was twelve he bought a 6-inch (15-cm) reflecting telescope with money earned from his paper route and spent clear nights studying the sky from his family's backyard.

King graduated from the University of Illinois at Urbana-Champaign with a degree in German. In 1979, he moved to Duluth, Minnesota, and worked as a photographer and photo editor for the *Duluth News Tribune* until retiring in 2018. He teaches community education astronomy classes, lectures widely and writes the Astro Bob blog (astrobob.areavoices.com) about what's up in the night sky. He also writes for *Sky & Telescope* and Universe Today.

King is author of *Night Sky with the Naked Eye* (Page Street Publishing, 2016) and *Wonders of the Night Sky You Must See Before You Die* (Page Street Publishing, 2018). King is married with two adult daughters and routinely sacrifices sleep for stars. But now that he's retired, he gets to take more naps.

INDEX